To Theresa

twenty-first century ire

GW00703474

Dying for Water

Sean McDonagh

VERITAS

Published 2003 by
Veritas Publications
7/8 Lower Abbey Street
Dublin 1
Ireland

Email: publications@veritas.ie
Website: www.veritas.ie

ISBN 185390 616 6

A catalogue for this book is available from the British Library.

Veritas books are printed on paper made from the wood pulp of managed forests. For every tree felled, at least one tree is planted, thereby renewing natural resources.

Design by Pierce Design
Printed by Betaprint Ltd

For Philomena Meaney, a gracious person
and a wonderful carer

Oh, waters of life! Full of noble virtues
You are a beacon of light,
Divine and pure,
envelop me in your majestic
tides and hold me secure.

– Rig Veda

Oh! Come to the water all who are thirsty;
Though you have no money, come

– Isaiah 55:1

Praise to You, my Lord through Sister Water,
Which is very useful and humble and precious
and chaste.

– Francis of Assisi, The Canticle of Brother Sun.

Whoever drinks this water will get thirsty again;
But anyone who drinks the water that I shall give
will never thirst again.
The water that I shall give
will turn into a spring inside them, welling up to eternal life.

1 John 4:13b – 14

Contents

Introduction

This short book looks at the importance of water for all the creatures on planet earth. However, it also details how humans have abused fresh water and the oceans for the past sixty years. Enormous damage has been done and the challenges now facing this generation are awesome. Water-borne diseases are responsible for 80 per cent of all illnessess and deaths in Third World countries. At present, one third of the world's 6.4 billion people live in water-stressed areas. Unless we introduce major changes, that figure will rise to two-thirds of the population by the year 2032.

This book also attempts to make solid connections between this contemporary ecological crisis and the Christian faith. Strange as it may seem, links between ecology and faith have only been forged in recent years.

I entered the seminary at St Columban's, Dalgan Park, Navan in 1962. The 1960s was a very exciting time for the Catholic Church as the new thinking emanating from the Second Vatican Council began to filter into seminaries and the wider Church community. It seemed that the Catholic Church, after distancing itself for centuries from the modern world, now wished to enter into dialogue with it. Renewed emphasis on the scriptures, the Church as the People of God, ecumenism, social justice, global poverty and human rights were high on the Church's agenda. Unfortunately, however, Vatican II did not appear to be aware that the global environment was deteriorating, despite the fact that Rachel Carson's pioneering book *Silent Spring* had been published in April 1962, six months before the Council began.

I learned nothing about the links between ecology and religion during my seven years in the seminary, despite studying the bible, the incarnation and the sacraments during that time. Sad to report that this is still the situation in many seminaries and departments of theology in universities.

My education in ecology began in the mid-1970s. I had finished a degree in anthropology in the US and was teaching anthropology and linguistics at the Mindanao State University in Mindanao, Philippines. I was asked to study a tribal people called the T'boli who lived in the rainforest of South Cotabato on the island of Mindanao in the Philippines. There I saw at first hand the appalling damage that the timber logging craze of the previous two decades had wreaked on the forest, rivers, lakes and even on the coral reefs. The health of the people was also being affected as the destruction of the forest meant the loss of medicinal plants and many food sources.

Once the trees were cleared and the area burned, the exposed topsoil began to silt up rivers and beaches, especially after heavy monsoon rains. Watching the rivers in raging floods sweeping the soils down to the sea was like watching blood haemorrhage from a body.

But was the ecological crisis confined to tropical areas like the Philippines as some commentators maintained at the time? I spent many nights in the T'boli hills reading environmental literature in an attempt to inform myself about these complex issues. I learned that tropical deforestation, global warming, acid rain, the extinction of species, depletion of the ozone layer, soil erosion and the chemical pollution of land and water were global problems. As I read this catalogue of greed, plunder and devastation I felt something of what the poet

Gerard Manley Hopkins captured so vividly in his poem 'Binsey Poplars', written after revisiting a favourite river bank where the trees had been chopped down:

> O if we but knew what we do
> When we delve or hew –
> Hack and rack the growing green!
> Since country is so tender
> To touch, her being so slender,
> That, like this sleek and seeing ball
> But a prick will make no eye at all,
> Where we, even where we mean
> To mend her we end her,
> When we hue and delve;
> After-comers cannot guess the beauty been...

Given the magnitude of the destruction that was taking place, and the fact that it will impoverish each succeeding generation of humans, I was amazed to find that the Catholic Church, either in the Philippines or globally, had nothing to say on the ecological crisis. It was all the more difficult to understand because the Catholic Church had a credible record on human rights abuses and social justice issues, and nature played a central role in Catholic sacramental theology.

I resolved that my work as a missionary would have a strong environmental thrust. Like many other missionaries, I was involved in education both at the primary and secondary level. I tried to include environmental studies in the curriculum with special emphasis on local issues like deforestation. Each school developed its organic garden and a compost pit. Parents were encouraged to revive their backyard gardens and cultivate them in an organic way. This meant that we had to find agricultural advisors who were trained in organic methods or were willing to learn them. Most existing graduates in agriculture were hooked on conventional, petrochemical, intensive agriculture.

Our primary health care programmes were very focused on increasing food production to boost nutrition and ensuring that the water sources were protected against contamination by animal or human waste. The high levels of infant mortality in the T'boli hills in the late 1970s was due to malnutrition and water-borne diseases like gastro-enteritis.

Creation figured very much in our religious programmes and liturgies. This was not difficult as the T'boli were very sensitive to the presence of 'spirits' in the world of nature. We tried to present the gospel of Christ as good news for humans and all creation. We incorporated planting and harvesting rituals into our ceremonies. The various creation symbols in the Easter vigil liturgy – water, light and darkness, fire and sexuality – allowed for all kinds of creativity in music, symbol and dance.

On the wider front I began to talk and write about the ecological crisis that I saw unfolding in Mindanao and right around the world, including Ireland. In the early 1980s I sent a manuscript of my first book *To Care For the Earth* to almost twenty different publishers in Ireland, Britain and the United States. All responded with a polite refusal saying that they did not feel there was a market for ecological theology. In fact, for most of the publishers, including those who were publishing radical liberation theology, these two worlds, like oil and water, did not mix. Finally, in 1984, a British publisher agreed to publish the book if market research convinced them that it would sell.

After the publication of *To Care For the Earth* I began to lobby Filipino bishops to publish a pastoral letter on the plight of the Philippine environment. The Episcopal Conference gave the green light and eventually *What is Happening to our Beautiful Land?* saw the light of day in February 1988. This was the first pastoral letter from any Bishops' Conference devoted exclusively to environmental problems. I was honoured to have played a part in the drafting.

Slowly the environment has begun to appear on the radar screen of Catholic teaching. The first papal document devoted exclusively to ecology appeared on 1 January 1990. It was entitled *Peace with God the Creator, Peace with all Creation*. In #15 the Pope stated that: 'Christians, in particular, know that their responsibility within creation and their duty towards nature and the Creator are an essential part of their faith. As a result, they are conscious of a vast field of ecumenical and inter-religious cooperation opening before them.'

More recently the Pope has become even more alarmed. On 17 January 2002, in an address on the destructive interaction between humans and the rest of creation, he said:

> However, if one looks at the regions of our planet, one realizes immediately that humanity has disappointed the divine expectation. Above all, in our time, man has unhesitatingly devastated wooded plains and valleys, polluted the waters, deformed the earth's habitat, made the air unbreathable, upset the hydrogeological and atmospheric systems, blighted green spaces, implemented uncontrolled forms of industrialization, humiliating – to use an image of Dante Alighieri ('Paradiso' XXII, 151) – the earth, that flower-bed that is our dwelling.
>
> It is necessary, therefore, to stimulate and sustain the 'ecological conversion', which over these last decades has made humanity more sensitive when facing the catastrophe toward which it was moving. Man is no longer 'minister' of the Creator. However, as an autonomous despot, he is beginning to understand that he must finally stop before the abyss.[1]

By any standard this is extraordinary language from a pope. He is aware that we do face a potential catastrophe and that none of the major institutions of our world – educational, economic, political, industrial, religious or media – have addressed the issue in a way that is commensurate with the seriousness of the threats facing the planet. After the World Summit on Sustainable Development in Johannesburg in August 2002, an editorial in the *New Scientist* (7 September 2002) wondered whether the countries of the world should set up a World Environment Organisation that has extensive powers to fine countries or corporations that wreck rainforests or renege on greenhouse gas commitments signed up to in the Kyoto Protocol. 'Cynics would say that governments would never give up such powers to an international body. But look at what they gave up to the World Trade Organisation in the interests of free trade'.[2] It seems that free trade and corporate profits are more important to our political leaders than the health of the planet.

Bishops' Conferences around the world have published various documents on the environment. Some are excellent, like *A New Earth – The Environmental Challenge* published by the Australian Hierarchy in 2002. They also set up Catholic Earthcare Australia to advise the Bishops on environmental questions. The European Bishops' Conference has held a number of consultations on environmental issues. The most recent, in Wroclaw, Poland, in May 2003, addressed the theme 'Formation for Responsibility for Creation and Sustainable Development'. This body has also encouraged pastoral initiatives on ecological themes like 'Day of Birds', 'Day of Trees' and 'Day of Bread'.

While the Irish Catholic Bishops' Conference has still to produce a document on the environment, the first pastoral letter devoted exclusively to ecology, entitled *The Whole of Creation is Groaning,* was published by the Archbishop of Cashel and Emly, Dr Dermot Clifford, in Lent 2003. Creation theology, while very evident in the religious education programmes in schools at both the primary and secondary level, is sadly lacking at third-level.

During the past twenty years I have published five books and countless articles on ecology, justice and religion. All of these have looked at ecological and justice issues from the perspective of the gospel of Jesus. The topics covered include global warming, tropical deforestation and waste management, and how global institutions like the World Bank, the International Monetary Fund and the World Trade Organisation have promoted economic policies that have impoverished people and destroyed the earth.

Our religious tradition should help us to rise to these challenges. Water plays a central role in the Bible and in our Christian faith. Preserving and protecting fresh water and the oceans must be seen as part of our Christian responsibility to care for God's earth.

I would like to thank Toner Quinn, my editor at Veritas, for his encouragement and helpful suggestions. A world of thanks also to all my colleagues here in Dalgan Park, especially Fr Oliver Kennedy and Fr Pat Connaughton. Finally, I would like to thank my sister Máire for proofreading the text.

The Water of Life

The United Nations General Assembly Resolution 55/196 declared 2003 as the International Year of Fresh Water. It is easy to understand why the UN is so concerned about fresh water. At this moment in history it is clear to any researcher that the human community is facing a global water crisis. The cold statistics are as follows:

- 1.2 billion people, about one third of the world's population, have no access to clean water.
- 2.5 billion have no sanitation and, if we continue with our business-as-usual attitude, two-thirds of the world will not have sufficient water in thirty years time.

These figures are taken from the United Nations Environment Programme (UNEP) report, *Global Environmental Outlook*. The report is based on the work of over one thousand scientists and was prepared for the Summit on Sustainable Development which met in August 2002 in Johannesburg.[3]

Human activity is polluting water in rivers, aquifers, lakes and the oceans around the world. The situation is extremely serious and poised to get worse unless concerted action is taken at local, national and global levels. In many areas of the world humans are using water at a rate beyond which it can be replenished, thus endangering the hydrological cycle.

Fresh water

The *Global Environmental Outlook* report tells us that 97.5 per cent of the world's water is in the oceans. Only 2.5 per cent of the waters of the world are fresh water and much of that is locked up in ice and snow in Antarctica, Greenland and the Arctic. Around 1 per cent of the fresh water is available – which amounts to 0.01 per cent of the world's water.

During the twentieth century the human population tripled. Water consumption on the other hand jumped seven-fold. Worldwide the demand for water is doubling every twenty-one years. By 2020 water use is expected to grow by 40 per cent. The major factors that account for the increased use of water are population growth, industrialisation and, especially, irrigation agriculture. It will probably come as a surprise to many people to learn that it takes 400,000 litres of water to manufacture one car, 2,000 litres to produce a glass of brandy, and 42,500 litres to produce one kilo of beef. [4]

To date, in order to meet the human demand for water, over 60 per cent of the world's largest 227 rivers have been dammed. In 1996 the United Nations estimated that humans consume about 54 per cent of the accessible freshwater in rivers, lakes and aquifers. As human population levels rise, the amount of available fresh water consumed by humans could jump another 20 per cent by the year 2025. [5]

Inequitable water distribution

It is estimated that there are about 12,600 cubic kilometres of water available for humans. The distribution is quite uneven. Canada receives twenty-six times more, per capita, than Mexico. Sixty per cent of the world's population live in Asia, yet it receives only 36 per cent of the world's fresh water. Of the fresh water available, 25 per cent is used by industry and 70 per cent by agriculture. [6]

In many First World countries water is readily available from taps, and some wealthy individuals have luxury swimming pools attached to their houses. The recommended basic amount of water needed by each person is around 50 litres per day. Four litres are needed for drinking and cooking and 25 litres are required to maintain personal hygiene. The rest goes into producing food. [7] The average person in the United States uses 600 litres of domestic urban water per day. Europeans use about 250 to 300 litres, while people in sub-

Saharan Africa use ten to twenty litres per day.[8] Almost 50 per cent of the water used by First World people is flushed down the toilet. Less than 10 per cent is used for drinking or cooking. The most wasteful aspect of our present toilet system is that the water in the cistern has been treated to drinking water standards. This water, which has undergone expensive chemical treatment and has been pumped for miles from the reservoir, is polluted with a single flush. Then the negative cycle begins again as the polluted water is pumped to sewage treatment plants.

In Mindanao, where I worked as a Columban missionary, many people, usually women and children, had to walk for miles to get their daily supply of water. Even then they had to use it sparingly. The same is true for vast numbers of people in Africa and Latin America.

The cost of water varies a lot. At the high end of the scale, if you live in Tanzania, you will pay 5.7 per cent of your daily wages on water. In Pakistan it costs 1.1 per cent of the daily wage. In Britain the average yearly cost to a household for water and sewage services ranges from £197 (Thames Water) to £327 (South West Water). There, a person pays 0.013 per cent of their wages on water, while people on the far side of the Atlantic, in the United States, pay as little as 0.006 per cent of their wages on water.[9] *Global Environment Outlook* estimates that it will cost $30 billion a year until 2015 in order to bring clean water to the poor of our world. If one includes proper sanitation the cost jumps to US$180 billion. This is still less than what the 2003 war in Iraq cost. The occupation of Iraq is costing the US $3 billion dollars each month.

Archbishop Renato R. Martino represented the Holy See at the Third World Water Forum in Kyoto, Japan, in March 2003. He argued that:

> The water concerns of the poor must become the concern of all in a perspective of solidarity. This solidarity is a firm and persevering determination to

commit oneself to the common good, to the good of
all and of each individual. It presupposes the effort for
a more just social order and requires a preferential
attention to the situation of the poor. The same duty
of solidarity that rests on individuals exists also for
nations; advanced nations have a very heavy
obligation to help developing nations.[10]

In Ireland drinking water is free for the ordinary citizen who
gets water from a municipal supply. It is the only country in
the EU or OECD not to have water charges. That could change
very quickly, however, despite the controversy that water
charges caused in the mid-1980s. Ironically, people in Ireland
now pay a fortune for bottled water. In 2002 we drank 110
million litres of bottled water.[11] Globally the bottled water
market reached the 100 billion bottle figure in 2002.[12]

In Britain the mineral water trade grew by 10 per cent in
2002. This is five times faster than the rate of growth for the
economy as a whole. This phenomenal growth was achieved at
a time when British water is improving by leaps and bounds.
The 2003 report of the Water Inspectorate found that 99.87 per
cent of the 2.9 million samples analysed met stringent UK and
EU purity standards. The inspectorate pointed out that the
water from the tap has never been cleaner and that it had the
advantage of being fresh while the bottled water may have
been on a warm warehouse shelf for up to two years. He also
observed that bottled water is one thousand times more
expensive than drinking water from the tap and three to four
times more expensive again if one buys it in a restaurant.
According to the editor of the *Guardian* (10 July 2003), 'why
the public images of drinking water is so grotesquely at odds
with the underlying reality is difficult to explain'.[13]

Water and Health, Water and Peace

Water is essential for all living organisms including humans. We each need to consume about four litres of water per day. At the most, we can survive for three days without water. Unfortunately the water many people drink each day, though it may slake their thirst, is not always potable. In the mid-1990s the UN Environment Programme listed twenty-five countries with poor access to clean water. Nineteen countries on that list are in Africa.

Water-borne diseases are responsible for 80 per cent of illnesses and deaths in poor countries. As a result, two million people, mainly in Africa, die needlessly each year from illnesses like diarrhoea, malaria, and other water-borne diseases.[14] This is the equivalent of ten jumbo jets crashing each day. One of the saddest statistics of all is that about 6,000 children die each day from water-borne diseases like gastro-enteritis. This figure is almost twice the number of people killed in the 11 September 2001 attacks on the World Trade Centre and the Pentagon which led to massive retaliation in Afghanistan and huge resources being poured into the war on terror. Unfortunately, there is no crusade to provide clean water for everyone on the planet, despite the high death toll across much of the Third World.

Bangladesh has plenty of water, over 80 inches of rain a year. Even so, finding potable water is often difficult. In recent decades villagers have installed hand pumps, so that they would not have to drink stagnant surface water that might contain animal or human faeces. Tragically, many of the deep wells were contaminated with naturally occurring arsenic. A study carried out by the British Geological Survey (BGS) in Bangladesh found that at least 1.5 million of 6 to 11 million tube wells in the county are contaminated with concentrations of arsenic in excess of the World Health Organisation's standard for drinking water. In the next few years, many areas of Bangladesh could face an epidemic of arsenic-related diseases, including cancer of the internal organs. Bangladesh urgently needs technical help in order to identify the areas of the country where groundwater is heavily contaminated with arsenic.

In 2003 a Bangladeshi inventor, Professor Fakhrul Islam, launched a new filter which is designed to extract arsenic and lead from tube wells. The filter, which contains a mixture of crushed bricks and ferrous sulphate heated together, was launched at a conference on arsenic poisoning in the US in 2003. Field tests, to date, have proved very successful. The filter costs $3 and can supply enough drinking water for a family of four. The United Nations is promoting the filter, but Bangladesh now needs the resources to spread the filter across the country so that millions of people will be spared arsenic poisoning which can be so painful and debilitating.[15]

Problems with groundwater

Until very recently humans relied mainly on streams, rivers and lakes for their water needs. According to *Global Environment Outlook,* two billion people now depend on groundwater (water found below the surface of the land, i.e. in soil and in the pores between sedimentary rock) for their water needs. It can take hundreds of years for water to percolate down into the aquifers. In recent decades groundwater has been depleted and polluted on an unprecedented scale. Currently we are pumping about 200 billion cubic metres (one cubic metre = 908 litres) more than can be recharged.[16] The demands on groundwater from agriculture, manicured lawns and 560,000 swimming pools have increased in California and the Southern Great Plains of the United States. By 2020, experts predict major water shortages.[17]

In India, 'the number of tube wells used to draw groundwater has surged from 3,000 in 1960 to 6 million in

1990'.[18] In most continents aquifers are being emptied ten times faster than they are being refilled. In certain areas of China, the water table is falling at a rate of 1.5 metres each year and it is estimated that over four hundred of the cities in Northern China face water shortages.[19] The aquifer beneath the Huang-Huai-Hai plain in Eastern China supplies drinking water for 160 million people.

In Libya, Colonel Qadjafi has initiated an enormous engineering project to draw water from the aquifer beneath the Sahara desert and pump it 3,500 kilometres to the coast to provide water for agriculture. The projected cost of the scheme is $32 billion. The ecological costs are also huge. Firstly, since there is no rain in the Sahara the aquifer will not be replenished. No one is sure, at the present, how long the supply will last. The estimates are between fifteen and fifty years. If it is closer to the first figure the cost of this water for irrigation could be as high as $10,000 per hectare. Secondly, there will inevitably be subsidence in the desert and the Nile waters could seep into the empty aquifer, endangering the whole population of Egypt.[20]

Because groundwater takes so long to recycle, polluting it is much more serious than polluting rivers. Today we are doing that at an alarming rate, forgetting that the average recycling period for groundwater is 1,400 years, as opposed to only twenty days for river water.[21] Organochlorines, human invented chlorine-based chemicals, are used widely around the world. The pesticide DDT, an organochlorine, is still contaminating groundwater in the US thirty years after it was banned.

The level of nitrates in groundwater has also increased in recent years.[22] The city of Ludhiana is home to over one million people in India's Punjab region. This is a most fertile area, known as the bread-basket of India. It is also home to many industries including textile factories, electroplating industries and metal foundries. The city is almost completely dependent on groundwater and this water is now highly polluted with cadmium, cyanide, lead and pesticides. A senior official at India's Central Groundwater Board laments that 'Ludhiana City's groundwater is just short of poison'.[23]

The chemical experiment

Unfortunately, the full consequences of today's chemical-dependent and waste-producing economies may not become apparent for another generation.[24] In December 1999, Dr John Peterson of the W. Alton Jones Foundation in the United States told a conference of scientists in Japan that:

> a hundred or more novel chemicals are swilling around in our bloodstream, chemicals which, before this century, were not found in human beings. It makes all of us, as well as our children and grandchildren, a walking experiment – one with completely unknown results.[25]

Some of these chemicals disrupt the endocrine system and therefore affect all aspects of human development from the embryo onwards. Many scientists are now convinced that the widespread use of organochlorines has led to an increase in testicular cancer and a decrease in sperm count in males. They also affect other mammals. Scientists who have studied the beluga whales in the St Lawrence Seaway have found high levels of toxic compounds like polychlorinated biphenyls (PCBs) in their blubber.[26]

Scientists and environmental non-government organisations (NGOs) are worried about the long-term consequence of these chemicals and have demanded that substances that are suspected of acting as endocrine blockers and that accumulate in human tissue should be banned. The Women's Environmental Network in Britain claim that between 1 and 8 per cent of babies in Britain have suffered slight nervous-system damage and memory loss as a result of being exposed to dioxins and PCBs.[27] Because the chemical and pharmaceutical industries are so central to modern economies, and so powerful, governments have been slow to investigate, regulate and ban these substances. Organic waste from sewage and animal waste is also a huge problem. It is often forgotten that in a country like the US, for example, farm animals produce 130 times more waste than the entire human population. Much of this ends up in streams and rivers and some seeps into the groundwater. The level of nitrates in groundwater has also increased in recent years.[28] High concentrations of nitrates in drinking water can cause infant methemoglobinemia or the blue-baby syndrome.

Water and world peace

Today the water situation in the Middle East and North Africa is precarious. North Eastern China, Western and Southern India, Pakistan, much of South America and countries in Central America like Mexico, face water scarcity. Somewhere in the region of 260 rivers flow through two countries or more. Only a handful of these countries have signed treaties regulating their respective access to the water.[29] As a result, competition between adjacent countries for access to water resources is causing friction and this could lead to outright hostilities in the future.

The conflict between Israel and Palestine is one of the running sores of our time. We know it has the potential to destabilise both the Middle East and the world. We seldom hear of what might be called the 'water dimension' of the conflict. In the occupied territories, Jewish Israeli communities consume seven times as much water, per capita, as the Palestinian Arabs.[30] Out of a potential annual flow of 1,250 million cubic metres (MCM) of water from the Jordan river, 565 MCM are utilised by Israel. The pumps which Palestinians used to draw water from the Jordan were destroyed by the Israelis after the 1967 war. There is also inequity and controversy over access to groundwater. The Western Aquifer System is a good example of this inequality. It has a safe annual yield of 362

MCM. Israelis now use 340 MCM, leaving a meagre 22 MCM to the Palestinian population.[31]

But the potential for water wars is not confined to Israel and Palestine. In March 2002 there was a stand-off between Malaysia and Singapore over the threat by Malaysia to cut water supplies to the island nation of Singapore. Lee Kuan Yew, Singapore's founding father, warned that any threat by Malaysia to cut water supplies to Singapore, even though Malaysian citizens are experiencing a severe drought, could lead to 'serious consequences'.[32]

The Tigris-Euphrates river valley is another potential hotspot. Turkey has spent $30 billion on dams and irrigation systems, forcing countries that are downstream, like Syria and Iraq, to curtail their water requirements. In 1989 Turkey threatened to cut water flow in the Euphrates because of Syria's support for Kurdish guerrillas.[33]

Until very recently Egypt, Ethiopia and other countries that share the Nile's waters seemed to be on a collision course over access to a fair share of the water. Almost sixty million people in Egypt depend on the Nile for their water, but the river rises elsewhere. About 85 per cent of the waters of the Nile come from rainfall in Ethiopia flowing as the Blue Nile into the Sudan. The rest comes from the White Nile with its headwaters at Lake Victoria in Tanzania. The Nile is the longest river in the world and supplies water to nine countries. Under the auspices of the United Nations those nine countries have recently reached agreement on equitable access to the Nile's waters.[34]

There is also the potential for tension between Pakistan and Afghanistan over sharing the waters of the Indus river. The Indus and its tributaries rise in the Indian Himalayas. On its journey to the sea it flows through Kashmir and into Pakistan. Pakistan relies on the Indus and its tributaries for crop irrigation and hydroelectricity. These rivers irrigate an area of Pakistan larger than England.

After years of bickering following independence and partition in 1947, India and Pakistan signed the Indus Water Treaty in 1960. This treaty ceded all the water in the eastern tributaries to India. In recent years the water in the western tributaries has been assigned to Pakistan. The on-going guerrilla war in Kashmir has come close to driving both countries to war in recent years and has put enormous strain on the treaty. In May 2002 prominent Indian politicians called on their government to scrap the treaty unless the government of Pakistan stops terrorists crossing over into Kashmir.[35]

All of these tensions are now exacerbated by the fact that the volume of water in the Indus has fallen in recent years, most probably because global warming has decreased the annual snowfall in the mountains. Less water will, in turn, lead to a food shortage as irrigation projects dry up. This will add tens of thousands to the seventeen million people who already live in poverty in Pakistan.

Conflict also looms on the Indian subcontinent, this time between India and Bangladesh. In December 2002 the Indian Prime Minister, Atai Behari Vajpayee, announced plans to pump water from the rivers in the north of the country through a series of canals to replenish the rivers in the South of India. The plan would:

> ... redraw the hydrological map of India, the waters from fourteen Himalayan tributaries of the Ganges and the Brahmaputra rivers in Northern India and Nepal, and transfer them south via a series of canals and pumping stations across the Vindhya mountains to replenish seventeen southern rivers including the Godavari, Krishna and Cauvery.[36]

If India implements this plan there is no doubt but that Bangladesh will resist it. According to the Ganges Treaty signed by both countries in 1996, India has agreed not to reduce water flow over the boundaries in future. The proposed plan would breach this agreement.

Elsewhere in Asia the Salween river is a source of tension between Thailand, Myanmar and China. Each of these countries has stated that they plan to dam the river in order that their own populations can benefit more from the water. The Mekong is a potential source of conflict between Cambodia, Laos, Vietnam and Thailand. The Mekong River Commission (MRC), with the various governments involved, will prepare a navigation strategy and programme by the end of 2003. The long-term goal of the strategy is to develop sustainable, effective and safe navigation on the Mekong and to increase the international trade opportunities for the mutual support of the member countries of the MRC.

Even in the United States of America there is disagreement among the various states about access to water. Right through the twentieth century there was conflict between Arizona and California about the allocation of the waters of the Colorado River. The Colorado River Compact in 1922 was supposed to settle the matter. It did not, and to this day, it continues to be an irritant between the two states. On New Year's Day 2003 the Federal Government ordered that three of the eight pumps that pump water from the Colorado River be turned off. The scarcity of water in this arid region could have major implications for agriculture in the United States. Six other states draw water from the Colorado River.[37]

It is no wonder that Ismail Serageldin, the World Bank's vice president for environmentally sustainable development, is on record as stating that many of the wars of the twentieth century were about oil, but wars of the twenty-first century will be about water. The prestigious business magazine *Fortune* (May 2000) concurs with this. It states that water promises to be to the twenty-first century what oil was to the twentieth.

Few doubt that Western need for oil was not a major contributory factor to both Gulf Wars – 1991 and 2003. One of the negative consequences of the 1991 Gulf War was the oil pollution in the desert. Sixty million barrels of oil spilled onto the deserts of Kuwait. They formed lakes covering 49 square kilometres. During the following years the oil has seeped down into the aquifer. Forty per cent of groundwater is now poisoned in a drought-prone country.[38]

Water and other creatures

Planet Earth supports millions of species, all of whom need water to survive. Today, one species – *homo sapiens* – now siphons off over half of the freshwater of the world and, as a consequence, endangers the survival of thousands of species. If human population levels grow as expected, the percentage of available water which humans will appropriate will jump 70 per cent by the year 2025.

Wetlands, which are essential for the hydrological cycle, have been devastated by human activity. Half of the bogs, swamps, fens, marches, estuaries and tidal flats have been destroyed in the past fifty years. These ecosystems store and purify fresh water by filtering out or diluting pollutants like heavy metals, phosphorus and nitrogen. Furthermore, the creatures that have thrived in these ecosystems, sometimes for millions of years, are now heavily stressed and in some situations face extinction. It is estimated that one fifth of the world's fresh water fish species – 2,000 out of 10,000 – are vulnerable and facing extinction. Mussels and mollusc species are also in danger. Over half the amphibians in North and South America and Europe are in decline. But the negative knock-on effect does not stop there. Over 1,000 species of birds, many of whom are dependent on wetlands, are facing extinction because of the loss of their habitat.[39] In January 2003 the administration of President George W. Bush, well-known for its anti-environmental policies, began legal proceedings that would remove obstacles to 'developing' millions of acres of isolated wetlands under the Clean Water Act.[40]

Water in Ireland, Britain and Australia

Ireland is blessed with a plentiful supply of rain. While the level of water pollution in Ireland has not reached that of Eastern Europe, there is no room for complacency. An editorial in the *Irish Times* in March 2000 stated that:

> Ireland's rivers have gone, in little more than a generation, from being almost pristine pure and clear to overblown imitations of open sewers and chemical drains.[41]

The editorial was in response to a preliminary report published by the *Three Rivers Project* which examined the water quality of the Liffey, the Boyne and the Suir. The judgement is somewhat exaggerated, but it does recognise that there has been a serious deterioration in water quality during the past forty years.

It is now accepted, even by the Environmental Protection Agency (EPA), that the quality of water in many Irish rivers and lakes has deteriorated in recent decades.[42] Fish kills, though less common than they used to be, still happen each summer. Kills have fallen from a high point of 112 in 1987 to 25 in 2001.[43] Fish kills result indirectly from the increased levels of phosphorus entering our rivers from a variety of sources. These include agricultural sources, sewage treatment plants, households, septic tanks and factories. The resultant algal bloom depletes the supply of oxygen and causes the fish and other aquatic life to die. What can also happen is that slurry, entering the water body, raises the biological activity in the water causing oxygen depletion in the short term. Slurry may also have other effects , as the run-off can be highly toxic.

According to the EPA's *Water Quality in Ireland 1998-2000* report, over 30 per cent of Irish rivers are polluted. This adds up to 4,006 kilometres of river channels countrywide. Seventeen per cent are moderately polluted, 12.4 per cent are slightly polluted and 0.8 per cent are seriously polluted.[44] Agriculture is suspected of being responsible for most of the recorded (300) instances of slight pollution with sewage accounting for the bulk of the remainder.[45]

Where drinking water in Ireland comes from surface or groundwater, the EPA found that overall compliance with drinking water directives stood at 94 per cent, with the public water supplies 93 per cent compliant in relation to iron, manganese and coliforms. The change in compliance for heavy metals and nitrates had improved over the 1999 levels, while the change in compliance has decreased in relation to aluminium, ammonium, coliforms, colour, manganese and fluoride. The EPA found that private water schemes are more susceptible to coliform contamination than public schemes, and official records point out that private group schemes are less well managed than public ones.

By 1 January 2004, Ireland is required to be in compliance with the EU Drinking Water Directive, 98/83/EC. The relevant Statutory Instrument is SI 439 of 2000. Dr Martin Knox, a water specialist, does not expect that Ireland will be in a position to comply with the Directive, mainly because of lack of investment in protecting water sources and adequate treatment, the lack of political will to enforce compliance and an inadequate infrastructure.[46] Agriculture, especially slurry spreading, is a significant culprit in surface water pollution. Experts like Dr Knox recommend that, like any other potential pollutant, farmers ought to be licensed in order to spread slurry. The licence would allow him/her to spread a fixed number of kilos per hectare depending on the nutrient content of the material. Farmers would be required to base their nutrient enrichment regime on an objective analysis of what is needed for a particular crop during a given year. Naturally farmers will need technical and financial support to be able to meet these goals and to work within what promises to be a more professional way of operating. Because the nutrient levels in soil would be known, farmers may not need to use artificial manures, thereby cutting the cost of crop production. A person with any knowledge of the farming community will know that these are heavy demands at a time when income from agriculture is falling. Recent statistics indicate that farm incomes were down 8.5 per cent in the year 2002. I do not know anyone in the environmental lobby who would wish to put small farmers out of business through saddling them with costly pollution control schemes. They realise that farmers play a crucial role in protecting the environment and also that, if small and medium size farms are eliminated, the ranches that replace them will do infinitely more damage to the natural environment. In addition, ranches will destroy the social and cultural fabric of rural communities and towns. Therefore, farmers will need incentives and well-targeted funding in order to be able to improve environmental controls on their farms. It is logical and cost-effective to invest in structures and policies that protect water sources instead of having to invest much more money to clean up polluted water.

The Three Rivers Project: Boyne, Suir and Liffey

In November 2002 the final report on the Three Rivers Project came up with findings similar to the EPA Report. The Three Rivers report established that agriculture accounted for almost 60 per cent of the phosphate pollution, 20 per cent was caused by urban sewage and quite an amount from septic tanks. Interestingly, from a Tipperary and Waterford perspective, while the quality of water in both the Boyne and Liffey had improved, the Suir has continued to deteriorate with almost 50 per cent of it slightly or moderately polluted.

The report called for a review of all existing discharge licences affecting the Suir catchment from South Laois to

Waterford. It wants waste-water treatment plants to be upgraded in towns and villages and it challenges the farming community to adopt best farm management practices, suggesting that each farm should be certified annually by an accredited planner.

Private Group Schemes

A report on the water quality in private group schemes was concerned in particular about the quality of Irish groundwater. It found that 38 per cent of the groundwater samples tested positive for faecal coliforms (E-coli) and animal and human sewage. The former chief executive of the Food Safety Authority, Dr Patrick Wall, warned that Ireland faced a potential fatal outbreak of E-coli 0157. Eleven people died in Canada after drinking water contaminated with E-coli 0157.[47] The report on the quality of water in private schemes found high levels of total coliforms and faecal coliforms in wells in both North and South Tipperary.[48] The source of the pollution is often from slurry spreading, improperly situated septic tanks and over-intensive grazing. Even holy wells are being polluted. In November 2002 coliform bacteria were found in a holy well in Kilkenny town which is also a popular source of drinking water.[49]

Aluminium levels in some water samples in Kilkenny are one hundred times above the EU limits and there are also concentrations of iron which are twenty-two times above the limits. The aluminium levels are particularly worrying as the World Health Organisation (WHO) has recognised the link between high aluminium levels and brain lesions, and with Alzheimer's disease.[50] In May 2003, the Environmental Protection Agency was criticised in a report for the European Parliament's Petitions Committee for not using more independent sources than the local authority in its study of the quality of Kilkenny's water supply. The authors, Mr Felipe Camisón Asencio MEP and Ms Margot Kessler MEP, also felt that the EPA's conclusion were based on an inadequate number of samples.

The MEPs also criticised Galway County Council. Local petitioners from Carraroe had complained that untreated sewage was contaminating the drinking water supply. The petitioners complained that Galway County Council continued to delay taking any remedial action. But these are not isolated cases. 'Given what most people think they know about Ireland, one would imagine that water would be the least of the country's problems,' the MEPs said.[51] However, judging from the number of petitions that are received from Irish citizens, clean, potable water is a major problem in Ireland. More worrying still is that, despite the rise in the incidence of contaminated water, the number of tests carried out by the State has, in fact, fallen. In a sharp rebuke to the EPA the report claimed that:

> The EPA has a greater tendency to defend the position of local authorities which supply it with most of its information than its own integrity as an independent agency which is supposed to be acting to safeguard the population.[52]

The EPA rejected the criticism, claiming that they are satisfied with the overall quality of water and that they have also monitored the data. It is significant that despite their robust defence of their record they did not state whether or not they were using independent data in this new monitoring climate.

Robert Pocock, who is the spokesperson on fluoridation for the environmental organisation VOICE, claims that 'routine overdosing of public water with flouride and aluminium is widespread, with only seven counties fully compliant with the legal requirements.'[53] According to the EPA Drinking Water Report 2001, all but two of the forty-two sanitary authorities in Ireland add aluminium sulphate to clarify their drinking water. Many counties exceed the maximum 0.2mg per litre. He goes on to point out that the

US EPA has warned that fluoride increases the toxicity of aluminium in drinking water. When 1mg fluosilic acid – the fluoride chemical used in Ireland – is added to every litre, fluoro-aluminium complexes are formed. These complexes increase the intake of aluminium by the body. In one study reviewed by the US EPA it was found that 80 per cent of rats which were exposed to only 1/2mg of aluminium died within a year from kidney or brain damage.

A group of international scientists, including Dr Paul Connett, Professor of Chemistry at St Lawrence University in Canton, New York, criticised the findings of the Fluoridation Forum Report in 2002. They stated that:

> Because the 'Fluoridation Forum' has failed to demonstrate either the efficacy of fluoridation or its safety, or convincingly demonstrate that this is an issue over which the government still has the right to overrule individual's rights to 'informed consent' to medication, it is time for the Irish government to halt this practice.[54]

Eutrophication and algal bloom

An EU-funded study conducted by Europe's leading experts on cyanobacteria found that 20 per cent of the lakes that it studied contained algal blooms which were caused by cyanobacteria.[55] An editorial in the *Irish Times* in January 2000 stated that:

> Cyanobacteria are among the most lethal substances known and can pose a severe threat to drinking water. They have been responsible for many animal deaths, and cause acute effects in humans, including skin conditions [and] vomiting.

The report acknowledges that 'it considered that the level of toxicity encountered substantially underestimated the true position'. The excessive use of phosphate fertilisers, inappropriate slurry spreading and phosphates in household detergents leads to eutrophication. This happens when excessive nutrients enter a body of water like a lake and patches of microscopic plants called algae proliferate because of the abundance of nutrients. Two things happen as a result. The green slime prevents sunlight from reaching plants in the water, and at the same time the decaying algae uses up the dissolved oxygen in the water, thus suffocating fish and killing other aquatic organisms. The stench from the water at this point is very unpleasant. It also means that the water is not suitable for human or animal use unless it is treated, and even the recreational use of the waterway is undermined.

Eutrophication is often seen as a local problem affecting areas like the Great Lakes in the US in the 1950s and 1960s and Irish midland lakes in the 1980s. The causes are attributed to industrialised agriculture and the domestic use of phosphates. Now scientists are beginning to see it as a worldwide problem for freshwater bodies. Phosphorus is one of the nutrients that plants need to grow. Until very recently the life-cycle of a phosphorus molecule took millions of years to complete. First it is released from rock by erosion to be incorporated into soils. The molecule moves on into plants and animals. On their death the molecule is returned to the soil often to be carried through rivers into the sea. In geological time this sediment is transformed into rock and pushed up to form land and begin the cycle once again.

Technologies developed in the twentieth century, especially in agriculture, have speeded up this process enormously. We now mine phosphorus in places like North Africa, Florida and Russia and ship it around the world to be used mainly as fertiliser. Writing in the environmental journal *The Ecologist*, Elena Bennet and Steve Carpenter state:

> Globally, we estimate that the annual accumulation of phosphorus in the earth's freshwater and terrestrial ecosystems has almost quadrupled from around 3.5 terragrams per year before humans began mining and

farming on a large scale, to around 13 terragrams per year now.[56]

As a result of this increase, eutrophication in lakes and estuaries is increasing. In the Gulf of Mexico, there is an area which is referred to as a 'dead' zone. Agricultural run-offs from the bread-basket of the US are carried by the Mississippi River right down into the Gulf. As a result there has been a 'massive die-off in ocean species'.[57]

With the agricultural changes in Ireland from the 1960s onwards many Irish lakes have been polluted. *The Water Quality in Ireland 1988 – 2000* document reports that:

> 85 per cent of the 304 lakes examined have been assigned to a trophic category (oligotrophic or mesotrophic) indicating satisfactory water quality and suggesting a low probability of pollution. The water quality of the remaining 44 lakes examined was less than satisfactory with the likelihood of significant impairment of their beneficial use.[58]

This is especially true of shallow midland lakes like Lough Sheelin which succumbed to eutrophication in the 1970s. The introduction and enforcement of a Slurry Management Programme in the 1980s led to a marked improvement in the water quality.[59] The water quality deteriorated once again in the mid-1990s. The architect Paul Burke-Kennedy believes that the wild brown trout of Lough Sheelin are facing extinction. It is obvious why this is happening. Too much slurry is still being spread on unsuitable land. Piggery operators have been granted Integrated Pollution Control (IPC) licences by the EPA. One wonders whether the EPA is aggressively monitoring how the piggery operators are getting rid of their slurry. Another contributory factor in the eutrophicalion of Lough Sheelin was the discharge of phosphate from the Mullingar sewage treatment plant. This has now been rectified by the installation of tertiary treatment of waste water. Paul Burke-Kennedy believes that:

The remedy is clear; it lies with political will and courage. But time is precious and no-one seems to understand the urgency.[60]

This may involve treating the slurry in anaerobic digesters.[61] Tourism is badly affected by pollution. Years ago Lough Sheelin was full of anglers during mayfly season. Now only a few people come as there are very few trout in the lake. When the trout stock was surveyed in 1978 for the Shannon Regional Fisheries Board the CPUE (catch per unit of effort) for trout was 5.0. By 2002 the CPUE had fallen to 0.6 and it is still dropping. There are twenty-seven intensive pig-rearing units in the surrounding area of Cavan. All the piggeries were granted planning permission, but many got this retrospectively through the retention route. While deliberate dumping is no longer taking place, slurry is still being spread at times of the year when the land cannot absorb it and it is finding its way to the lake with disastrous consequences for the trout and the one-time flourishing tourist industry.

At the national level a millennium study by the EPA found that fertilisers in agriculture were responsible for 75 per cent of plant nutrient input into Irish waterways.[62] It is no wonder that in the mid 1980s even large lakes like Lough Derg on the Shannon and Lough Ree were also in trouble. A 1999 report from a Catchment Monitoring and Systems Group found that 'Lough Derg is moderately eutrophic (moderately polluted), while Lough Ree is strongly eutrophic (seriously polluted)'.[63]

In April 2002 two environmental groups, Save Our Lough Derg (SOLD) and Save Our Lough Ree (SOLR) claimed that sewage from 'one-off' houses, agricultural run-off and inadequate municipal sewage treatment plants represent the biggest threat to water quality in the Shannon.[64] Two years after the 1999 Report, a new study released in July 2001 found that 35 per cent of the waterway of the Shannon system was significantly polluted, though this was down by 10 per cent on a previous report. The report found that there has been a slight improvement in the quality of Irish lake water – 304 lakes were tested; 85.6 per cent were found to be unpolluted; 3.9 per cent were slightly polluted; 7.2 per cent were moderately polluted and 3.3 per cent were seriously polluted (or hyper-eutrophic).[65]

Though there have been improvements in the wake of the completion of a number of sewage treatment plants, upgraded farmyard storage facilities and adherence to the phosphorous spreading guidelines, the situation on the Shannon is still critical. One of the main farming organisations, the Irish Farmers Association (IFA) drew up its own nutrient management code.[66] Other measures have also helped. The Rural Environmental Protection Scheme (REPS) encourages less intensive agriculture, requiring each farmer to follow a strict nutrient management regime. Forty-five thousand farmers had signed up to REPS by the year 2000.

But still much more needs to be done. On 24 March 2003, the *Irish Times* carried a photo on page two of a warning sign on Lough Ree on the Shannon. The warning read as follows:

> Toxic algae may be present on or near the foreshore in this area. The level of algal toxins present can be dangerous to both humans and animals. If algae or scum is clearly visible, do not swim or allow animals to drink the water. Roscommon Co. Council.

Similar signs have appeared on the Tipperary and Clare shores of Lough Derg in recent years. One was erected in Dromineer (a popular bathing area on Lough Derg) in July 2003.

Many First World countries, including Ireland, are attempting to deal with the excessive use of phosphorus, especially in agriculture. Most of the schemes are voluntary. In the Netherlands restrictions on the use of phosphorus are more stringent and are backed up by legislation. Farmers are subject to manure quotas for each crop and acre of their land. If they exceed this quota they are fined. If we wish to protect the waters of Ireland a similar approach may be needed here.

Most efforts at phosphorus reduction have been local, regional or national. Lake Leman (also known as Lake Geneva) and Lake Constance were heavily polluted in the early 1980s. Commissions for the protection of these lakes were set up with the aim of monitoring and reducing the amount of phosphorous that was allowed enter the lakes and the rivers that fed them. Twenty years later the water quality in both these lakes is acceptable for drinking and for bathing.

While these successes ought to spur more remedial action in Ireland, it is important to recognise that eutrophication is a global problem often caused by run-offs from hundreds of miles away so there is a pressing need to tackle the issue at a global level. This will involve a serious balance between the twin goals of producing sufficient food for a growing world population while at the same time protecting the waters of the world.

Irish Government's inaction on water challenged by EU

Given our geographical location on the edge of the Atlantic ocean and our frequent rain, it is almost unbelievable that the Irish government is in breach of the EU Water Directive. In 1998 a number of environmental organisations in Ireland complained to the European Commission that the Irish government was not fulfilling its obligations under the EU Water Directive. In April 2002 the Advocate General of the EU rejected the defence put forward by the Irish government for not ensuring that Irish drinking water was up to EU standards.

The Advocate General, Mr M. A. Tizzano, recognised that some progress had been made in Ireland, but his judgement concluded that, 'although appreciable, these initiatives are not capable of sustaining the obligations ... nor can they be used in justification for its infringement'. He sustained three complaints. First, Ireland had not ensured compliance with the Directive in relation to the public water mains network and certain parts of that network in particular. Second, Ireland had not ensured compliance concerning faecal coliforms in relation to certain group water schemes. Third, Ireland had no binding national legislation which it could apply to group water schemes.[67]

The judgement was finally handed down in November 2002. The European Court of Justice ruled against Ireland under the EU Drinking Water Directive. In relation to the contamination of group water schemes in rural areas of the country the court found that 'the argument that Ireland had taken steps to improve the quality of its water intended for human consumption cannot be accepted'.[68] The court stated that each one of the 145,000 houses in group water schemes all over the country must benefit from the Directive.

Many environmentalists who are concerned about Ireland's deteriorating water sources have lost faith in local or national attempts to stop pollution. They feel that local authorities have been unwilling to face down powerful lobby groups in agriculture, industry or in the services and, while the establishment of the EPA in the early 1990s helped, there are still major problems with the enforcement of environmental legislation. Environmentalists are now hoping that the EU's Water Directive will force the Irish government to protect rivers and lakes or face major fines. Dr Martin Knox believes that strict enforcement will involve a major cultural change in the way directives are monitored and enforced. There are two pieces of legislation in relation to waste water dating from 1977 and 1990 which are not being enforced. The 2003 Environmental

Protection Bill will focus the minds of local authorities on the requirements for better monitoring and control according to Dr Knox. Fines for non-compliance will be much higher. He also believes that payment for domestic water will introduce discipline into the way consumers use water. In saying this, Dr Knox is well aware that provisions will have to be made for those who are unable to pay. According to him, this ought not to invalidate the principle that we need to pay for water, so that the resources are there to protect it, purify it and enforce the relevant Directives. Dr Knox feels that, in the light of the technical and resources constraints on local authorities, there is a need for a National Water Authority that would oversee the treatment and distribution of drinking water. Such an organisation will need to be properly resourced and priority must be given to the training of technical staff and enforcement initiatives rather than to the bureaucratic and organisational side of any new structure. In this regard, lessons can be learned from examining the allocation of budgets within the health services in Ireland. The expenditure on health doubled in a five year period from the mid-1990s onwards even though the service to the consumer does not appear to have improved appreciably. Many feel that the increased expenditure was swallowed up in administrative staff rather than focused on the front line personnel of nurses, doctors and consultants in the various areas of health. The same mistake must not be made in the management of a precious resource.

I have concentrated on the poor quality of Irish freshwater. It is also worth mentioning that the quality of Irish coastal waters has deteriorated during the past ten years because of poor or non-existent sewage facilities and agriculture run-offs. As a result, shell-fish produced in Clew Bay and Kenmare Bay were downgraded from category A to category B in July 2003. This will mean huge losses for the industry which has marketed Irish shell-fish on the basis of its pristine quality.

The final paragraph of the summary in *Water Quality in Ireland 1998-2000* claims that 'water quality in Ireland for the 1998-2000 period is generally satisfactory in comparison with most other European counties'[69] Not everyone would agree with that statement. Most people would agree with the hope and sentiments expressed in the rest of the paragraph.

The fresh approach now being adopted as a result of the EU Water Framework Directive gives hope that the long-term chronic water quality problems can be tackled successfully. Sustainability requires that the present generation hands on to the next precious resources such as viable salmon and trout populations in rivers and lakes. In so doing an improved aquatic environment will benefit everyone – for example, in terms of increased tourism revenues, improved angling and amenity value and also improved sale of Irish milk and meat products taking advantage of a 'green' competitive advantage. A deteriorating aquatic environment will drive away tourists and anglers, increase the cost of water treatment, depress agricultural sales and deprive the next generation of it's birthright, i.e. access to a high quality Irish environment.[70]

Many NGOs believe that the *Water Framework Directive* (WFD) could be the instrument to help clean up Irish waters.[71] Between 2003 and 2009, seven River Basin Management Plans, similar to the Three Rivers Project above, will be written for Ireland. River Basin committees have been set up with NGOs representatives on each committee. These River Basin Committees will have a vital role in ensuring that enforcement agencies such as the local authorities and the Environmental Protection Agency do deliver real action on the ground. They are also ideally placed to raise awareness and give a lead in providing relevant education about the condition of water in any particular area.

A position paper produced by the environmental organisation VOICE in cooperation with other environmental NGOs entitled *The Water Framework Directive – issues and opportunities* (2003) stated that, while local authorities have been given the lead responsibility on Ireland's River Basin Projects, they are currently poorly equipped, in terms of finances and internal expertise, for the job. In mid-2003 local authorities have no staff allocated to the ecological and environmental aspects of the WFD. They only have technical and engineering expertise. Local authorities work on a county level rather than on a catchment area basis. Often responsibilities relating to water quality, pollution control and management issues are passed from one organisation to the next and no action is taken. This highlights the need for much greater co-ordination among statutory bodies, agencies, government departments and local authorities and indeed it might be the role for the water authority recommended by Martin Knox. Finally, if the WFD is to be successful in meeting its targets, monitoring must be carried out, not by local authorities themselves but by independent experts.

Water in Britain

In Britain there are also problems protecting fresh water. Over 4,000 rivers in Britain are polluted. During the 1960s and 1970s a lot of attention was devoted to improving the quality of water in Britain's rivers and canals, but improvement programmes came to a standstill during the 1980s. As a result the levels of industrial and agricultural pollution in many important rivers increased dramatically according to the National Rivers Authority (NRA).

An EU Report on water in Europe conducted in 1998 found that Britain has more cities and towns dumping raw sewage into the sea than any other country in the European Union. Among the eleven towns and cities cited are Torbay, Hastings, Hull, and Dundee. The sewage treatment in other major cities cited like Edinburgh, Glasgow and Liverpool is considered to be inadequate. The EU has threatened to bring the British government to the European Court of Justice where it could be forced to pay a sizable fine unless it cleans up its act quickly. British officials claim that improvements have taken place since then.[72]

The Mersey, once considered one of the most polluted rivers in the world, has still not been fully cleaned up despite efforts in recent years to treat the raw sewage that used to be discharged into the river. There are hopeful signs, however. In 1976, seventy-seven tons of mercury were flushed into the estuary. By 1989 this had fallen to four tons. In addition, people can draw some comfort from the efforts that have been made in recent decades to clean up the Thames.

There was an improvement in bathing water quality in Britain by 2002. Ninety-seven per cent of Britain's beaches were considered suitable for bathing in 2002; 341 beaches reached the top rating in 2002, a jump of 66 beaches on the year 2000. Every region of Britain, except Northern Ireland, showed an improvement in 2001. The reason for the improvement was attributed to upgraded sewage facilities and calm weather, as in previous years severe storms had breached sewage facilities.[73]

Australia

Global Environment Outlook reports that in Australia the quality of water in many inland waterways has declined due to human activities within catchments. Sediments, nutrients and toxic materials, as well as excessive growth of aquatic weeds, have affected aquatic ecosystems. Even though Australia is the driest continent on the planet, it is estimated that each Australian uses one million litres of fresh water annually.[74] This is almost twice what people use in the US.

In September 2002 the Australian Catholic Bishops Conference published a social justice statement entitled *A New Earth: The Environmental Challenge*. The document expresses the bishops' concern about the state of rivers in the country.

The health of our rivers is a national issue. River stress is a major issue in the Murray-Darling Basin, and for all the south-east coastal river systems of Victoria and New South Wales, the agricultural regions of south and central coastal Queensland, the south-west of Western Australia and northern Tasmania. We can still enjoy and protect the great rivers of the continent's far north which remain mostly pristine, free and wild.

The health of the Murray-Darling epitomises the ecological crisis. This once great waterway now surrenders 80 per cent of its flow for human consumption. Since European settlements between 12 and 15 billion trees have been lost from the Basin. This river system, which is a major artery of Australia's agriculture, is exhausted and dying. Because of water removal for irrigation, the river at times has not the strength to reach the sea.

As Columban Father Charles Rue puts it:

Farming in Australia means working within the limiting factor of the availability of water rather than arrogantly following European farming practices. Railing against drought is really a confession that we have overstocked or over-cropped the land in inappropriate ways.[75]

Privatising water

At present, the global population is 6.6 billion. This is expected to rise to 8 or 9 billion by the year 2025. The demands for clean, potable water will become acute in the next few decades. Increased demand and shrinking supplies are a sure way to make huge profits. Transnational Corporations (TNCs) see the potential in supplying water and sewage services globally. Their PR companies are trying to get governments, local authorities and even citizens to see water as a commodity rather than a right. Their slogans argue that unless people pay for water they will not value it.

The potential for a lucrative market is obvious when one looks at the statistics globally. In 2001 only 5 per cent of the world's water needs was provided by water corporations. That means that 95 per cent are still being provided by governments, local authorities or communities. The corporate world sees huge potential for expansion and the likely profits are enormous if they could get control of 50 per cent of the market. The corporations that are already in the business are doing extremely well. Since 1990 the world's three largest water companies – Suez, Vivendi (both French) and Thames' Water have expanded into every continent on the planet. The International Consortium of Investigative Journalists (ICIJ) reported that Vivendi's earnings from water-related revenue was $5 billion in 1990. By 2002 it had risen to $12 billion.[76]

The business magazine *Fortune* estimates that the water market is a £280 billion-a-year-industry. Others estimate the annual revenue as anything between $400 billion and $3 trillion. It is an assured market. One can live without computers or gasoline and a host of other things. But one cannot live without water.

The corporations see that there is ample room for expansion, especially as international agencies like the World Bank, the International Monetary Fund (IMF) and the World Trade Organisation (WTO) have championed the privatisation of water in many Third World countries.[77] The ICIJ reports that in poor countries the World Bank has used its financial leverage to force governments to privatise their water utilities in exchange for loans. The ICIJ investigation found that in the Philippines the World Bank had been advising the leaders to 'commercialise' their utilities as part of an overall bank policy of privatisation and free-market economics.[78] The ICIJ study of 276 World Bank water supply loans from 1990 to 2002 found that 30 per cent required privatisation, a majority of these in the past five years.[79]

The World Bank justifies its support for privatisation of water utilities by arguing that private companies can circumvent much of the bureaucratic morass and corruption that one finds in the public sector in many countries. The ICIJ study counters this argument by pointing out that private companies have invested little of their own capital in water utilities relying primarily on the World Bank and other international financial institutions to help cover the cost of repairing and expanding utility networks. The ICIJ study claims that if the Bank applied the same energy and money in improving local utilities while allowing them to retain control of their water systems, the local utility would actually perform better than the private company.

In Manila for example, Maynilad Water, which is 40 per cent owned by Suez, announced in December 2002 that it was pulling out of its twenty-five year contract and abandoning a waterworks serving 6.5 million people. The company was unable to raise capital to meet contract demands and, as an independent study commissioned by the local regulator indicates, contracted work to affiliate companies. Maynilad is before an arbitration court seeking $337 million from the Philippine government as reimbursement for what it claims it invested in the project. The Philippine government maintains that the company is owed only a fraction of that amount. Maynilad wants to transfer to the government debts totaling $530 million.

The rhetoric that promotes privatising water is all positive. The corporations will be efficient, reliable and caring. ICIJ found that in practice the situation was not that rosy.

> The investigation showed that while these companies claimed to be 'passionate, caring and reliable', as one company states, they can be ruthless players who constantly push for higher rate increases, frequently fail to meet their commitments and abandon a water works if they are not making enough money. As in South Africa, the water companies are pillars of a user-

pay policy that imposes high rates with little concern for peoples' ability to pay. These rates are then enforced by water cut-offs despite the serious dangers to peoples' health that these actions create.[80]

The water corporations also cherry-pick their projects. They want to the get their hands on the lucrative side of the market supplying water to the rich and middle classes. They are not interested in supplying even a small percentage of the water needs of the poor in the mega-cities they have targeted around the world. Within a year of privatising water the charges increase significantly. Those who cannot pay are simply disconnected from the system and left to fend for themselves. The poor have then to pay exorbitant prices, often for polluted water.

Many non-government organisations (NGOs) and development and environmental organisations saw the third World Water Forum (WWC) in Kyoto, 16-23 March 2003, as an effort by the corporate world to become central players in providing water services globally at a cost. The conference was not organised by the United Nations but by the World Water Council. This is a private think-tank with a strong corporate and pro-privatisation bias.

The corporate vision, unfortunately often supported by governments, is that the so-called free market is the best mechanism for delivering water because it recognises that water is an important commodity. During the final session activists accused the members of the World Panel of Financing Water Infrastructure of 'sacrificing the poor for profit'.[81] The activists chanted slogans like 'No profits from water', 'Water for life, not for war', and 'Water is a human right'.

That latter point was emphasised by Amnesty International's statement on the Kyoto meeting.

> **Amnesty International expressed deep disappointment at the failure of the international community to**

recognise the human right of every person on the planet to clean water in the final ministerial Declaration at the World Water Forum in Kyoto.

Amnesty's statement further asserted that the declaration is a backward step, as a UN expert committee has recently affirmed the human right to water.[82] It is important to remember that the Forum is not the United Nations and therefore does not have international legitimacy.

The California-based International Rivers Network circulated a flyer at the conference that began:

> Who's Behind the World Water Forum? A brief guide to the world water mafia. A web of think tanks, corporations, agencies and lobby groups is attempting to control the global discourse on water problems and solutions...

It went on

> WWC is a lobby group heavily weighted with engineering and construction companies, dam-building state agencies and water supply corporations'.[83]

For Amnesty and many other environmental organisations and religious groups, water is a public and social good and ought not to be primarily regulated by so-called market forces. If water does become a mere commodity then it will be accessible only to the rich and the middle classes. The poor will be excluded with appalling consequences for hundreds of millions of people.

The address by Archbishop Renato R. Martino, the Vatican Representative to the Third World Water Forum, made a similar point and was critical of the rush to privatise water. He stated that

Water by its very nature cannot be treated as a mere commodity among other commodities. Catholic social thought has always stressed that the defence and preservation of certain common goods, such as the natural and human environments, cannot be safeguarded simply by market forces, since they touch on fundamental human needs which escape market logic (cf. Centesimus Annus, 40).

Water has traditionally been a State responsibility in most countries and viewed as a public good. Governments worldwide, for diverse political and social considerations, may indeed often provide large subsidies to insulate water users from the true cost of water provisions. Being at the service of its citizens, the State is the steward of the people's resources which it must administer with a view to the common good.

At the same time, in the interest of achieving more efficient sustainable water services, private sector involvement in water management is growing. It has, however, proved to be extremely difficult to establish the right balance of public-private partnership and serious errors have been committed. At times individual enterprises attained almost monopoly powers over public goods. A prerequisite for effective privatisation is that it be set within a clear legislative framework which allows governments to ensure that private interventions do in fact protect the public interest.[84]

Later on in his speech Archbishop Martino makes it clear that the Holy See views the right to clean water as a basic human right. It acknowledges that this right is only mentioned explicitly in the Convention on the Right of the Child. The Archbishop went on to support those people who are now campaigning to formally include the right to water as a basic human right.[85]

It has not been all smooth sailing for TNCs in the water business. There have been problems. In 1999 the World Bank, which has championed privatisation globally, threatened to withhold $600 million in debt relief unless the government of Bolivia privatised its water facilities. In response the government handed over the inadequate water system of Cochabamba to a company called Aguas Del Tunari. This company is a joint venture between Bechtel, a huge US construction company, and United Utilities, a British company. Within a month water bills jumped by 35 per cent. Demonstrators drawn from unions, student and environmental groups took to the streets in Cochambamba, Boliva, and the city ground to a halt. A US trained captain in the Bolivian army shot dead a seventeen-year-old student, Victor Hugo Daza. This enraged the demonstrators and many more people were hurt in the on-going violence. After months of rioting and with six people dead, Aguas del Turani officials fled the city. At an emergency meeting government officials rescinded the contract and declared that Aguas del Turani had abandoned their 40 year contract worth a whopping $2.5 billion. The water facility has now reverted to public ownership, but Betchel are suing the Bolivian government for $25 million to cover their development costs in Cochabamba.[86] Curtis Runyan, writing in *WorldWatch* magazine, believes that

> ... privatisation schemes around the world have resulted in drastic rate increases, significant job cuts, fewer environmental safeguards, dropped conservation initiatives, and halted services to poor or remote communities.[87]

In Ireland, business interests also see the potential for a lucrative water market. Already the National Toll Roads' Celtic Anglian subsidiary has secured a €220 million water treatment plant in Ringsend in Dublin. Their partner is the publicly quoted company Anglian Water of Britain. A subsidiary of the French water company Ondeo called Degremont has been given the contact for a water drainage contract in Cork worth

95 million euro. In the National Development Plan which covers the period 2000-2006, 3.8 billion euro has been set aside for water and sewage facilities. The breakdown of that figure is that 1.657 billion euro has been earmarked for waste water schemes, a further 579 million euro for water supply sources, 862 million euro for rehabilitation and management of the existing infrastructure and, finally, 702 million euro for water and sewage schemes to support economic activity. Naturally there is huge pressure from business interests and the European Commission to privatise the delivery of the water they consume and, effectively, to force people to pay for it.

From what we have seen so far it is clear that at a global level humans have not been able to achieve an equitable and sustainable use of fresh water in recent decades. If the present trend of population growth in Third World countries and rising living standards in First World countries continue there will be massive pressures on water resources in the next few decades. It will mean that by 2032 two-thirds of the world's population will be living in water-stressed areas unless major changes happen in the next ten years.

People like Dr Peter Gleick, director of the Pacific Institute for Studies in Development and Security in Oakland, California, are hopeful that humans will avoid this bleak future. He argues that even now we possess all the necessary technology to achieve sustainable water use by the mid-twenty-first century. It will mean abandoning many of the traditional big dam solutions and opting for 'a mix of traditional and innovative small-scale water supply systems, locally managed and environmentally sensitive'.[88]

On the industrial front he points out that Japan has reduced its water use by a quarter since the 1970s, even though industrial output has risen significantly in the intervening years. In agriculture we saw that there are irrigation technologies which are between 50 to 95 per cent more efficient than current large scale irrigation projects, especially in Israel.

The ordinary citizen can make a real difference by the choices he or she makes. In the United States all new toilets sold since January 1994 have, by law, been high-efficiency, low-flow installations. This has cut down by an astonishing 70 per cent the amount of water needed to flush millions of toilets in the United States. Gleick also calls for mind-set changes in two areas. Firstly, it will be necessary to wean engineers and planners away from large scale, grandiose water schemes. Secondly, governments will have to stop subsidising the profligate waste of water.[89]

In a statement to the Third Water Forum in Kyoto, Peter Gleick disputed whether it is necessary to spend $80 to $100 billion to solve the current water crisis globally. He maintained that $10 to $20 billion would go a long way towards addressing the problem, provided the money was spent on community based projects and not on massive, expensive and problematic dams. 'Instead of our current bias towards large-scale centralised water projects, we must invest aggressively in community-scale water projects that bring basic water and sanitation services to those who need it most.'[90] Ten or even twenty billion dollars to provide clean water for everyone in the world is small change when compared with military spending. In the year 2000, $829 billion was spent globally on buying arms. Almost $20 billion of that was spent by Third World countries.

Patrick McCully, writing in the *New Internationalist*, believes that there is still a bias towards large dams and reservoirs in the minds of planners.

> There has never been a fair playing field when dams have been compared with their alternatives. Corruption, and the power of the big dam lobby, both in government and corporations, has meant that feasibility studies for new dams have regularly underestimated the cost and exaggerated the benefits. If assessments of options for water and energy needs were made

comprehensive, transparent and participatory, very few large dams would make the grade.[91]

This tallies with my own experience while working for over a decade with the T'boli people in south east Mindanao in the Philippines. Our water and sanitation programmes were very local, low-tech and very participatory. We found that there was no need for dams and large piping schemes. The most important aspect of rural water and sanitation schemes is that people are educated to look after the system themselves.

Water and Agriculture, Industry and Tourism

Water is essential for growing food. Agriculture accounts for 80 per cent of water consumption worldwide. With the increase in human population in recent decades there has been an increase in agricultural activity and irrigation, but it would appear that in many parts of the world our use of water in agriculture is unsustainable.

Farmers have been drawing out so much water from wells in northern China that the water table is sinking by more than a metre per year. Wells near Beijing are now more than 1,000 metres deep. The most pressing current problem is finding water for agriculture. Rice and wheat are staple foods in different parts of China and they require huge quantities of water. It takes 2,000 tons of water to grow one ton of rice and 1,000 tons of water for a ton of wheat or maize.

Plants and animals are thirsty. They need a lot of water. Given the need for food and the recent rainfall patterns in China the government is toying with the idea of diverting water from the Yangtze to supply water eight hundred miles away in drought-stricken Northern China where, as recently as 2002, three hundred cities were experiencing water shortages. Many observers fear that such a diversion could have disastrous effects on the environment.[92]

Water table levels are falling in China with resulting water shortages that many commentators believe will not be solved by the controversial Three Gorges Dam project, which I discuss later in this chapter. Even the United States is beginning to have trouble in this context. The Ogallala aquifer spans eight states in the United States. It stretches from South Dakota to north-west Texas. It supplies about one quarter of the irrigated farms in the United States. It too is being drawn down at unsustainable rates.[93]

Pakistan faces the same crisis situation with regard to water. The Indus is the main source of water for human consumption and agriculture. This mighty river begins in Tibet and flows for over 1,000 miles through Kashmir and Afghanistan before entering Pakistan. Before the river was pock-marked with dams and irrigation schemes, and spilled out into the ocean through a 600,000 hectare delta, this area was remarkably rich in marine life and paddy rice agriculture. The rice yield was so abundant that apart from being able to feed its own citizens Pakistan was able to export rice to the Gulf States. Due to pressure on the river from irrigation canals and dams upstream this has all changed in recent years. Rice production has continued to fall each year for the past ten years. In 2001 it fell by 369,000 hectares. Life is no better for fishermen. In delta towns like Keri Bundar, fishing families have been forced to move in search of fresh water on numerous occasions as salt water has contaminated the remaining waters of the Indus.

Engineering schemes to harness the Indus for agriculture began in the Punjab during the British regime in the early 1890s. Successive irrigation and power schemes in the 1930s, 1950s and 1960s have reduced the flow of water to the delta and given rise to the present crisis. The lack of fresh water and seepage of salt water into the delta is also having a devastating impact on biodiversity. The once rich fishing grounds have been devastated.[94]

Sandra Postel, the director of the Global Water Policy Project in Amherst, Massachusetts contends that even with our current technologies, farmers could cut down their water usage by one quarter and industry could save over 90 per cent if it recycled cooling water.[95] New irrigation technologies, especially the drip irrigation developed in Israel, delivers minute amounts of water directly to the plant roots through pipes buried alongside the plants. The system is 95 per cent efficient in delivering almost all the water to the plants and, of course, it cuts down hugely on the water used in irrigation through the traditional methods.

Dams[96]

Humans have dammed most of the major rivers of the world during the twentieth century in order to generate power or provide water for human consumption and agriculture. In 1900 there were no dams larger than 15 metres high. By the middle of the century that figure had jumped to 5,000. At the end of the century the number of major dams had reached almost 45,000.

At first sight and from a distance many might feel that dams have played an important role in providing energy and water for irrigation and human consumption. It is easy to see why dams were promoted by multilateral agencies like the World Bank from the 1960s until the mid-1990s. The negative side of dam construction has only begun to get an airing in official circles in recent years. Dams create great human and ecological problems.

The World Commission on Dams has estimated that between forty and eighty million people have been displaced by dams in the twentieth century;[97] 130,000 people alone were displaced by the Aswan dam in Egypt. Many of these people have never been properly resettled on suitable land and so their quality of life has continued to deteriorate. The same happened in Brazil. Those who suffered most from having their lands inundated were the indigenous people and small farmers.

Vast areas of land have been flooded in constructing dams in the past five decades. The reservoir on the Volta dam in Ghana flooded an area the size of the Lebanon. By 1981, almost 120,000 square miles of land worldwide – the size of Italy – had been flooded. Much of these lands were fertile areas or forested. The Balbina dam in Brazil flooded 890 square miles of tropical forest. The damming of the Narmada river in India has been one of the most contentious building programmes in recent years. It is the fifth longest river in India, 1,312 kilometres long. When all the dams on the river are constructed it is estimated that over 100,000 people will have been displaced. Many of these are tribal people and forest dwellers. The dams have also inundated some of the richest agricultural land in India.

Large dams and irrigation programmes have become ideal locations for spreading waterborne diseases like malaria. Some of the villages near lake Nasser, created by the Aswan dam, have now 100 per cent infection rate for bilharzia.[98]

Dams have disrupted the valuable nutrient flow which was carried by rivers into flood plains, deltas and the oceans. Before the Aswan dam was constructed the annual sardine catch in the eastern Mediterranean was 18,000 tons. A decade later the catch was reduced to 500 tons. There has been a significant loss in agriculture also. Before the construction of the Aswan dam 100 million tons of fertile sediment was carried on to farms along the Nile. The dam disrupted this flow and so, to keep up agricultural production, the supply of natural nutrients has to be made up by artificial fertilisers. In 1990 the cost to the Egyptian economy was over $100 million dollars per year. Because they are dammed in so many places some of the world's largest rivers like the Colorado in the US and the Yellow river in China are now reduced to a mere trickle as they enter the ocean.

Dams and perennial irrigation projects in many countries have led to water-logging and salinisation. The latter occurs when the delicate salt balance of soil is upset. This allows the salts to build up. Eventually the soil loses its structure, is unable to sustain plant life and thus becomes barren. Perennial irrigation schemes in hot climates where the land is never rested can become saline. Unless the land is well drained as well as irrigated, irrigation projects invariably cause the water-table to rise, thus bringing the salts to the surface. Salinisation afflicts at least one quarter of all irrigated lands.[99] The problems are particularly acute in Pakistan and Australia. In geologically active areas in Egpyt, China, India and Greece, the sheer weight of water has triggered earthquakes. An earthquake registering 5.6 on the Richter scale shook the area around the Aswan dam in Egypt in 1981. 230,000 people lost their lives in central China in 1975 when two large dams collapsed.

One of the largest construction projects in history is the Three River Gorges dams on the Yangtze in China. More than half a million people will have been moved from their homes by mid-2003. When it is completed by 2008, a further 700,000 people will have lost their homes according to official figures. Critics of the project claim that the more realistic figure for people losing their houses and land is closer to two million. Those who have opposed the dam in China have been subjected to human rights abuses. Abuses against those who have protested against proposed dams have also occured in many other countries, from the Philippines to Guatemala, where authoritarian regimes have attempted to force through mega-dam projects.[100]

The Three River Gorges dam is not the only water-engineering project in the pipeline in China. There is talk of bringing water from the wet south to the dry north. This would involve diverting 40 trillion litres of water per year from the Yangtze and moving it 800 miles north to provide water for agriculture and over 400 cities that now face water shortages. The project, first suggested by Mao Zedong in the 1950s, could take decades to complete and cost in excess of $24 billion dollars. Environmentalists fear that the project, if completed, could have devastating ecological consequences.[101] They point to what happened to the Aral Sea when the rivers that fed it were dammed and redirected.

In the late 1970s and early 1980s I was involved in a successful campaign to stop a major dam being built to combine Lake S'bu and Lake Siluton in the mountains of South Cotabato in the Philippines. The Catholic Church and other groups campaigned against the proposed dam as it would have inundated hundreds of hectares of rice land, and undermined the well-being of the T'boli people who have lived in the area for over 1,000 years. Campaigners were also successful in stopping the controversial five dams project on the Chico River in Luzon in the Philippines.

The Aral Sea

The destruction of the Aral Sea, located between Kazakhstan and Uzbekistan, is one of the great ecological disaster stories of the second half of the twentieth century. Until the 1960s it was the fourth largest inland sea in the world, comprising 65,500 square kilometres. In the 1960s a massive engineering project diverted its two feeder rivers – the Amu Darya and the Syr Darya – to irrigate cotton fields and rice paddies in Uzbekistan. The result is a catastrophe. The Aral sea is now only half its former size and it has lost three quarters of its volume. Wind storms blow the toxic sand from the exposed seabed on to villages, contaminating crops and exposing humans and animals to the poison.

The damage to fish and other forms of aquatic life has been horrendous. By 1990 only 38 of the 173 species that inhabited the sea had survived. Before the diversion there was a thriving fishing industry around the shores of the Aral Sea. Now there is no commercial fishing and many of the trawlers are abandoned on desert land almost a mile from the shore of the sea. The number of bird species has also fallen. There are only 285 species using the sea as compared with 500 species in 1960.

Plant life has also been decimated by the shrinkage and the increased salinity in the water. In 1960, 1,200 flowering species were counted in the vicinity of the sea. Today only a fraction survive and all of the twenty-nine species which were endemic to the area are extinct. The tugay forests which are unique to the area and once covered 13,000 square kilometres on the edge of the sea are now decimated. In truth what remains of the Aral Sea is a human-created hell on earth. This disaster ought to be a bleak reminder to us that we cannot continue to make war with the earth. Unless the human community learns to limit demand for water and is willing to share it with other creatures, what happened in the Aral Sea might be replicated worldwide in a few decades.[102]

Given these and other environmental catastrophes associated with irrigation it was assumed in the mid 1990s that the days of huge dams was over. 'The era of big dams is over' said Gordon Conway, vice-chancellor of the University of Sussex in Britain and a water consultant for the Ford Foundation and the World Bank. According to Conway, 'they don't work. Costs are increasing in real terms, the environmental and social consequences are considerable, and there are big questions about their efficiency'.[103]

My own fear is that large-scale engineering projects are back in vogue in 2003. Certainly, one of the few breakthroughs at the UN Conference on Sustainable Development in Johannesburg in August 2002 was in the area of water and sanitation. The Conference committed the international community to cutting the number of people not connected to potable water supplies by 50 per cent to 550 million by the year 2015 and to halve the number without proper sanitation to 1.2 billion by the same year. Such initiatives could save tens of thousands of lives. But, typical of the woolly nature of the Conference's thinking, there are no clear guidelines or effective strategies for achieving such laudable targets. In fact, according to the *New Scientist*, individual countries and even corporations are 'left free to pursue approaches to managing water that are either wasteful or damaging to the environment'.[104] According to Jamie Pittock, water director of the World Wildlife Fund (WWF), 'summit agreements to improve water will not work if natural sources of water are not conserved and water used more efficiently'. Transnational corporations would like to promote large building projects like dams and piping systems. The Summit played into the hands of the big building corporations according to Torkil Jonch-Clausen of the Global Water Partnership:

> This summit has reduced the debate on water supply to arguing about money and pipes. There is no discussion about managing our river systems. It is a step back to the 1980s, before Rio ... It is a prime example of how the development lobby (transnational corporations) have snatched back the sustainable agenda from environmentalists.[105]

Dams always produce less energy, generate less water and irrigate a great deal less than their proponents promise.

Water and industry

Industries also use huge quantities of water and often misuse water in a profligate way. In many industries water has become a solvent or cooling agent. Many people would be surprised to learn that it can take 25,000 gallons of water to produce a single motorcar. An average nuclear power station uses 30 million gallons per day. The US computer industry alone uses 400 billion gallons each day.[106]

Oil exploration, chemical and heavy metal pollutants from industry in the lower Volga have virtually destroyed the Caspian Sea, the largest inland body of water in the world, covering an area of 4,000 square miles. Drilling for oil began in 1874. By the early 1880s one hundred refineries were located in the area. The oil business continued under the communist regime. With the collapse of the USSR in the late 1980s, Western oil corporations began to invest in some of the countries that surround the Caspian Sea. As a result of the activities of the oil industry, part of the coast of Azerbaijan is severely polluted with phenols and oil products. For decades much of the heavy industry of the former USSR was located on the Volga. The river flushed heavy metals, chemical waste and untreated sewage into the Caspian Sea. This has led to miscarriages and stillbirths, and congenital deformities are very common. The coastline of Kazakhstan is equally polluted. Diseases like tuberculosis and blood diseases are four times higher near the coast than elsewhere in the country. Many of the drinking water supplies are polluted with oil spills.

Other species are faring no better. Fisheries have almost collapsed. In 1985, 30,000 tons of fish were taken while the annual catch in the late 1990s was down to 2,100 tons. The Azerebaijan Centre for the Protection of Birds claims that thousands of birds have died from oil contamination. Unless remedial action is taken very soon the Caspian will follow the Aral Sea as a testimony to human greed and destruction.

Tourism

International tourism, now the world's fastest growing industry, is putting huge pressure on springs, rivers and aquifers from Ibiza to Barbados. Of the seven underground springs in Ibiza, five have already been over-drilled. As a result salt water is seeping in, making the water unsuitable for drinking or agriculture.[107] Tourists in Africa who have access to daily showers and a continuous water supply will see women carrying water on their heads over huge distances and not make the connection between their profligate use of water and the consequent water shortages for the local population. The island of Barbados is a well-known holiday resort. There have been water shortages on the island since the 1950s. During the dry season many of the local population have no water supply because it is reserved for the half a million tourists who visit the island each year. Halfway around the world in Goa, luxury hotels require up to 60,000 gallons of fresh water each day. Much of this water is pumped from villages which have been left with only sandpipes. The over-drilling has affected the water table and, as a result, salt water is seeping into fresh water wells. Many tourists bound for Australia love to visit Ayers Rock. New hotels catering for the 400,000 annual visitors are putting pressure on the underground water sources and the local springs which were traditionally used by the Aboriginal people in the area.[108]

The World Wide Fund for nature (WWF) has estimated that a typical tourist in Spain uses 990 litres of water per day, compared with 250 litres used by local people. Most people will appreciate the demand that swimming pools make on water resources. Golf courses are equally demanding. In a hot, dry country an eighteen-hole golf course can consume as much water as a town of 10,000.[109] Spain's National Hydrological Plan (PHN) to carry water from the Ebro in north-east Spain to supply tourism and agriculture in the dry south-east led to demonstrations in many Spanish cities, including Madrid, Barcelona and Zaragoza, in the autumn of 2002. The wetlands in the Ebro's delta would be threatened by PHN's plans.[110]

Disappearing and polluted rivers

As we have seen already, dams have reduced mighty rivers like the Colorado to a trickle. The great Hoover Dam, built in the 1930s, broke all engineering records up to that point in time. This massive project became the template for engineers around the world to tame, and put to 'productive' use, the major rivers of the world during the following seventy years. Another important feature of Hoover was the height of the dam at 150 metres high. At the beginning of the twenty-first century there are more than one hundred such giant structures across the world, ranging from the Itaipu in Paraguay to the Bhakra Dam in India.

Dams bring benefits in energy generation, food production and human needs, but they also cause havoc to aquatic environments and biological diversity. Many species of fish are threatened with extinction once their habitat is disturbed.

Many of the great rivers of the world are heavily polluted. The Yangtze river in China is polluted with 40 million tons of industrial and sewage waste each day. In the 1990s, medical research among people who lived along the Hei river in central Anhui province found that mortality rates were 30 per cent higher than the national average. Many of the illnesses there are associated with poor water quality in both the Hei and Huaihe rivers. In 1995 the Chinese government closed 1,111 paper mills and 413 other industrial establishments in order to improve the quality of water.[111]

Elsewhere in Asia major rivers are also polluted. These include the Ganges (India), the Amu and Syr Darya (Central Asia). Rivers in Nepal's urban areas are contaminated and drinking water in Kathmandu is contaminated with coliform bacteria, iron, ammonia and other contaminants.[112] The Pasig river which runs through Manila is an open sewer. In Europe the Volga and the Ural rivers are contaminated with industrial effluent. Plans to clean them up have not been too successful.[113] Water pollution in Latin America and the Caribbean did not become a serious issue until the 1970s. Over the past thirty years there has been a significant decrease in the quality of surface water. The main culprits are agricultural run-offs and untreated urban and industrial sewage.

Destruction of the Oceans

More than 97 per cent of all the water on earth is sea water. *Our Common Future*[114] states that:

> In the earth's wheel of life, the oceans provide the balance. Covering over 70 per cent of the planet's surface, they play a critical role in maintaining its life-support systems, in moderating its climate, and in sustaining animal and plant life, including minute oxygen producing phytoplankton. They provide protein, transportation, energy, employment, recreation, social and cultural activities.[115]

The oceans have a very special place in the history of the universe, planet earth and the story of life. They emerged as the planet cooled and water vapour gathered into them. Water vapour or ice may have been found on other planets or comets, but earth is the only place in the Universe where we find running water and oceans. They have existed for almost four billion years. To us the oceans may seem ordinary, but we can truly appreciate their significance when we view them as the universe unfolding itself in a new way.

Furthermore, the oceans are the womb of life. Gradually, with the passage of time, more complex elements including amino acids and, finally, proteins, began to appear in the oceans. Then, suddenly, simple cells capable of reproducing themselves formed and life awakened on planet earth over three-and-a-half billion years ago. The microscopic green plants are called phytoplankton. They developed the wonderful technology called photosynthesis. This technology enables plants, using energy from the sun, to split hydrogen and oxygen and then to combine the hydrogen with carbon dioxide in order to create the organic compounds on which life depends. These compounds include proteins and sugars.

Oxygen, which is very important for humans, is released as a by-product of photosynthesis. This knowledge has been available for a long time. It is only recently, however, that scientists have discovered that phytoplankton remove as much carbon dioxide from the atmosphere as all the land-based plants and trees. In this way they reduce the percentage of carbon in the air and keep the planet much cooler than it would be if these tiny creatures were not providing this essential service.[116]

That the universe should burst forth into life in the oceans is an extraordinary event with massive possibilities for the future. Every one of the millions of species that have walked the earth, swam in the oceans, or have flown in the sky owes its existence to the first simple cells that emerged in the oceans. Yet in recent decades we have treated rivers, lakes and the oceans abominably.

During the UNESCO proclaimed International Year of the Ocean, in 1998, it emerged that the oceans are being over-fished and polluted at an unprecedented rate. Important areas of the oceans, close to continental shelves, are contaminated with human, agricultural, industrial and radioactive waste. Much of this is toxic and causes cancer. Because we have tended to treat the oceans as sewers, the Baltic, Mediterranean, Black, Caspian, Bering, Yellow and South China Sea have all been seriously damaged in recent decades. The waters of the Black Sea, once a flourishing eco-system, are now considered to be 90 per cent dead. Each year the Danube has an estimated 60,000 tons of phosphorus and 340,000 tons of inorganic nitrogen dumped into its waters. It has little chance of being flushed clean since it takes 167 years for the water from the Danube delta to reach the Mediterranean, and much longer to reach the Atlantic.

The *Sunday Times Magazine* (10 June 2001) carried a devastating report on the pollution of the Caspian Sea. It stated that 'if you are a child living by the Caspian Sea, venturing outside is a dangerous game to play'.[117] Some of the coastal areas of Azerbaijan are so polluted that 'still births, miscarriages and congenital deformities are common'.[118]

In short, we treat the oceans as sewers into which we pour our human, animal, industrial and even nuclear waste. According to *Global Environment Outlook* global sewage remains the largest source of contamination, by volume, of the marine and coastal environment, and coastal sewage discharges have increased in recent decades.[119] Calcutta and Bombay, respectively, pump 400 and 365 million metric tons of raw sewage into the ocean each year. In Pakistan, the figures for Karachi are that 175 million metric tons of sewage and industrial waste are dumped annually into the Arabian Sea. Elsewhere in Asia, China disposes of fifty to sixty million metric tons of untreated sewage into the coastal waters every day.[120]

Concerns about industrial pollution of the oceans were expressed at the time of the Stockholm Conference on Development and the Environment in 1972. The Minamata disease, caused by people eating mercury-contaminated seafood, had already happened. Even then sea birds had been affected by the organochloride DDT. As evidence of the toxicity of DDT and other poisonous chemicals mounted, measures were taken at that time to ban organochlorines like DDT and polychlorinated bipheniyls (PCB) in many countries. There are international conventions like the Oslo Dumping Convention designed to protect the oceans from pollution, but, unfortunately, these conventions have not been enforced in any effective way.

Another source of concern is the number of sea creatures that are affected each year by non-biodegradable litter. Tens of thousands of sea birds, turtles and marine mammals are killed each year by ingesting or becoming entangled in non-biodegradable litter.[121] I have a Bord Iascaight Mhara (Irish Sea Fisheries Board) poster entitled *Marine Debris, Biodegradation Time Lines* hanging on the wall of my office. According to their reckoning paper towels take two weeks to biodegrade. Cardboard can take up to three months to disintegrate. Tin cans and Styrofoam cups take 50 years. Monofilament fishing nets take as long as 600 years, while plastic bottles and disposable nappies take 450 years to biodegrade . Millions of these items are now polluting the sea. Tens of thousands of whales and dolphins have been injured and killed by these nets and plastic bags.

Over-fishing

On another front, over-fishing is depleting the oceans and leaving them barren. Many people feel that the oceans are so vast, and the variety of fish so abundant, that there will always be vast quantities of fish in the sea. We are now learning how false those assumptions are. According to a report by the UN Food and Agricultural Organisation (FAO) in 1995, over 70 per cent of the world's marine fish stocks are either 'fully-to-heavily exploited, over exploited, or slowly recovering'.[122] Populations of marlin, sword-fish, tuna and ray have crashed by 90 per cent in the past fifty years.[123] Every one of the fishery hotspots of the planet is now endangered. Even salmon are in decline. The fall off in pollack, cod and hake is more gradual. In the North Atlantic all high value species – herring, cod, etc. – are in decline. In the Mediterranean and Black Seas the fish stocks are completely exploited and the fish catches in the Black Sea have collapsed. In the Indian Ocean most important species are fully exploited. It is the same situation with fish stocks in the Central and Southern Pacific. Fish stocks in the Southern Atlantic are still in good shape, but recently long-range international fleets have moved in so the prospects for the future are not good. The situation in the southern oceans is also causing worry. The catches here are mainly krill and there is a danger that they are being over-exploited by Japanese boats. The problems with over-fishing were summed up eight years ago by Christopher Newton of the UN FAO when he said that 'the history of fishing is to postpone problems until you run out of fish, which is where we are now'.[124]

Most of the damage to the oceans has been done during the twentieth century. Fish catches have increased from three million tons at the beginning of the twentieth century to almost ninety million tons in 1989. Most of the increases happened after World War II when sonar and radar tracking technology, which had been developed for military purposes, were used to locate and catch fish. Fishing fleets now use spotter planes and satellite data to locate the fish. Furthermore, super-trawlers the size of a football field are built to accommodate nets thousands of feet long. In one single netting these boats can take up to 400 tons of fish.

The depletion is most notable in many of the world's most productive fishing grounds. These include the Grand Bank of Canada and New England. Cod fishing has collapsed in the North Sea. According to Jonathan Amos, BBC News Online (March 2002), if current over-fishing continues in the North Atlantic, trawlers could soon be left chasing jellyfish and even plankton to make 'fake' fish products. The claim comes in the wake of a comprehensive survey of fish stocks in the North Atlantic Ocean. Scientists estimate that across the region as a whole, fish numbers are now just one sixth of what they were one hundred years ago.

A BBC production team spent five years in the mid-1990s preparing a series of programmes on the world's oceans called *The Blue Planet*. One of the programmes scripted by a marine biologist, Martha Holmes, was entitled 'Deep Trouble'. It looked at the damage that humans are inflicting on the oceans at the moment. She concluded that the problem with present day fishing methods is not just that they are hoovering up everything in sight but that they are taking juveniles and destroying ocean habitats.

To illustrate the first charge she looked at what has happened to the bluefin tuna on the western side of the

Atlantic. These fish migrate from the Caribbean up the east coast of the US to Canada each year. Their routes are very predictable and during the last decade they have been exploited in an unsustainable way. The programme claims that, since the 1960s, the number of bluefin tuna has plummeted by 85 per cent on the western side of the Atlantic Ocean.

How did this happen? In the 1980s bluefin tuna was worth a few cents a pound. In the early 1990s entrepreneurs realised that if one could airlift them to Japan their value would increase dramatically. The price is now tens of dollars per pound. Holmes visited a fish market in Tokyo and found that fresh blue-fin tuna were fetching £60 a kilo. Sometimes, a dealer could expect £12,000 for one fish. Markets fluctuate wildly and the price can even go higher. Such an enormous price is guaranteed to wipe out the bluefin tuna in a few short years.

The shortage of fish in the North Atlantic is having an impact elsewhere. More and more fish are imported into Europe and North America from other parts of the world like West Africa and Asia, depriving poor communities there of a good source of protein. To date this replenishing of the stocks from other fisheries has hidden the crisis from the ordinary citizen and consumer. Marine scientists claim that only radical and comprehensive action can save the North Atlantic from an ocean-wide collapse in fish within the next ten to twenty years.

A realisation that things were really bad and poised to get worse began to dawn on people in March 2002. Britain woke up to the fact that one of their favourite fish species, the common skate, was almost extinct in the North Sea. The news was released at a conference of European environment ministers in Norway. The ministers called for tough measures, including closed areas, to protect what was left. In 2001 a concern about the fall in the number of cod led to a ten-week ban on fishing the species during the breeding season, but the measure was seen by the environment ministers as too little, too late. Even North Sea mackerel, which was once so abundant, is now classified as commercially extinct. Other popular fish like haddock, whiting and plaice are also endangered and need protection. There is now a call for a 50 per cent reduction in the European fishing fleet and a ban on certain kinds of fishing in order to protect what remains.[125]

Towards the end of 2002 more bad news appeared in the wake of advice from the International Council for the Exploration of the Seas (ICES). ICES is a Copenhagen-based intergovernmental scientific body which advises governments and the European Commission on fishery and environmental issues in the North Atlantic. The organisation has nineteen member states and each state is entitled to have two members on the Council. It recommends that a number of areas including the Irish Sea should be closed because of the alarming depletion of cod stocks.[126] The decline has been dramatic in recent years.

As recently as 1986, 10,000 tons of cod was landed from the Irish Sea each year. Out of that number Irish boats were responsible for 4000 tons. By 2001 the catch had fallen to 3,875 tons with 714 tons going to Irish fishermen. Efforts were made in the year 2000 to conserve cod stocks in the Irish Sea. These included a ten-week ban from mid-February plus the closure of the spawning grounds, between south County Down and the Isle of Man, and changes in the design of nets. But, according to ICES, more drastic steps need to be taken: the 'current state of cod stocks and the failure of past measures to bring fishing mortality down to rates that allow rebuilding means that more stringent action is required'.[127]

Further restrictions on cod fishing were introduced in 2003. In 2002 the quota for taking cod from the Irish Sea was set at 2,017 tons. In 2003 the quota has dropped 36 per cent to 1,284 tons. In the Atlantic off Donegal it is down by 61 per cent, from 1,035 tons in 2002 to 407 tons in 2003. The EU Fisheries Commissioner, Dr Franz Fischler, proposed new restrictions on cod fishing in the Irish Sea in May 2003. It is understandable that the fishing organisations are opposed to such cuts. The Commissioner pointed out that there were resources in the EU to compensate fishermen who are adversely affected by these new restrictions which are necessary if there is to be any stock recovery of cod in the Irish Sea.[128] Cod numbers in the Arctic Sea off the coast of Norway are also reduced. The same goes for the Canadian stock. The only cod stock that is faring reasonably well at the moment is the Iceland stock. If that drops there will be no Atlantic cod.

Quotas for plaice and sole are also reduced. The fishing industry in Ireland is encouraging people to switch to deep water species, like orange roughy and black scabbard.[129]

Atlantic Dawn

An Irishman has the dubious distinction of owning the biggest fishing boat in the world at the moment. It is called *Atlantic Dawn* and is owned by Kevin McHugh. According to *The Guardian* (20 February 2002) its nets are twice the size of the Millennium Dome. The boat has been registered as a merchant vessel by the Irish government in order to avoid European Union fishing restrictions which are aimed at preserving fish stocks. Mr McHugh managed to secure a private licence to fish in Mauritania's waters off the coast of West Africa for nine months of the year. Because the boat is 5,000 tons bigger than the size allowable, *Atlantic Dawn* was not able to qualify for a cash-for-access scheme whereby European boats are allowed to fish in Mauritania's waters.[130] The European Union has bought fishing rights in the area from the government of Mauritania which is one of the poorest countries in the world. Many in the environmental and development community insist that the EU agreement with Mauritania is one of total exploitation and destruction. In a short period it will deplete fish stocks and deprive local people of a traditional source of food and employment.

The UN environment programme has expressed reservations about the depletion of fish stocks by giant EU vessels. The main fish caught are sardinelle which is the staple diet of the local people. Depleting the stocks will add to malnutrition and hunger in the area. This is already beginning to happen. In 1997 Senegal produced 453,000 tons of fish. This was reduced to 330,000 in the year 2000.[131] A simple comparison between the catches of *Atlantic Dawn* and local boats captures the magnitude of the plunder. *Atlantic Dawn*, with a crew of 100 and automatic factory freezing facilities, will catch in a day what ten local fishing boats catch in a year. Its purse seine nets are 3,600 feet in circumference and 550 feet deep. The trawl nets are 1,200 feet in breadth and 96 feet in height. It is estimated that *Atlantic Dawn* catches in one month what 7,000 artisan fishermen would catch in a good year.[132]

Atlantic Dawn is able to process 400 tons of fish per day and has the capacity to store 7,000 tons of frozen fish. Beatrice Gomez of the Coalition for Fair Fisheries Agreements, which represents fishermen in Mauritania and environmental groups, said: 'We fear for the future of these stocks, and the EU allowing the Irish to register *Atlantic Dawn* spells disaster for local people. It should be stopped'. Mauritania is one of the poorest countries of the world and is highly indebted. During the past few years it has licensed its fishing grounds to factory fishing boats from Japan, China and the EU. As a result there has been a dramatic drop in catches during the past four years. When stocks are depleted, the factory ships move on elsewhere. Unfortunately, Mauritania's fishermen cannot move on. The fishing sector provides over 50 per cent of foreign exchange and provides in excess of 30,000 jobs.[133]

So in the era of the Celtic tiger, short term national economic gains take precedence over long term global destruction and the impoverishment of the poor in Mauritania.

Ocean catches peaked in 1989 as a direct result of such over-fishing. In 1998 they were down over 30 per cent despite improved gear, tracking and snaring technology. Daniel Pauly, the author of a new study on global fishing trends, predicts that 'if things go unchecked, we might end up with a marine junkyard dominated by plankton'.[134] Dishonesty and corruption are rife. Between 1986 and 1992 more than six times the quota for cod, flounder and redfish was taken from the Grand Bank off the Canadian coast. When Spanish ships were boarded by Canadian police in 1995, the Canadians found two sets of books on board. One recorded the true tonnage of the catch for the owners. The second set of books (with false, reduced figures) were meant for the authorities if the ships were challenged.

Martha Holmes acknowledges the discrepancy between fishermens' statistics about the state of particular species and the estimates that marine scientists come up with. It is little wonder that such discrepancies appear, given the scale of the profits involved. An Irish scientist, Mr Tim O'Higgins, is adamant that, 'attempts to conserve fish stocks should not rely on the "observer" system of monitoring catches'.[135] Mr O'Higgins is both a marine scientist and a graduate of NUI Galway and has worked as an observer in the North West Atlantic monitoring halibut in the Greenland area. He states that the 'responsibilities placed on the civilian observers takes no account of the reality on the ground'.[136] He recounts that when he had finished his monitoring duties the skipper of the vessel asked to see the figures and then exhorted him to change them. When he refused the skipper punished him by cancelling a rendezvous with an EU ship that was to collect him and bring him home.[137] This is an extraordinary accusation and if investigated and confirmed the skipper should be prosecuted and debarred from captaining a fishing vessel again. Only such resolute action will save our oceans.

Fishing at much deeper depths

With so much fishing pressure on surface waters fishermen are now going deeper and catching fish that our ancestors would never have known existed, never mind eaten. One can now find fish like the black scabbard at fish markets in Europe. This fish lives at depths of about 1,000 metres so until very recently it was outside the human catching potential.

Will we soon be doing irreparable damage to this and other species that live in the deep ocean? The answer is that we do not know. In fact we know very little about what happens in the ocean at this depth. It is only recently that scientists have been able to do research in deep waters. The fear is that many species will become extinct even before we have been able to study them at any depth. Humans have explored only 2 per cent of the deep ocean. We know far more about what exists on the surface of Mars than we do about what lies in the deep ocean. As Martha Holmes asserts in *The Blue Planet*, 'We could be in danger of losing a treasure trove of species before we have even discovered them'.

In the past few years marine scientists in Australia and New Zealand began to study the impact of fishing on a deep sea species called orange roughy. These can live at a depth of 1,000 to 1,500 metres. They also live to a ripe old age of 150 years. As a result, this means, of course, that they mature later than most fish we are familiar with, at about thirty years of age. If catches contain a high proportion of immature fish, then of course the species is doomed because the fish will not be able to reproduce.

Destruction of habitat

In the deep sea we are discovering very strange and wonderful animals. Their habitat, however, is under threat from the crude and destructive methods of fishing that are so often used. Firstly, the nets are so huge that they can scoop up everything on the sea bed, including corrals and sponges, some of which are hundreds of years old. The nets rip these creatures off the seabed, toss them into the nets and then discard them as by-catch. The final result is a totally destroyed marine environment that took hundreds of years to develop. A New

Zealand marine scientist interviewed by Martha Holmes on *The Blue Planet* explained that the best way for us land creatures to understand the destructive nature of deep sea fishing is to compare it with a similar way of getting food on land. If we took our cue from the fishermen in order to get some beef we would first hire a helicopter. Then we would trawl a huge net across the countryside. This process would capture a cow, the primary target of our endeavour. With the cow would come the dog, the car, the barnyard door, the farmer's wife and a whole lot of things we did not wish to get in the first place. Once the nets were wenched in we would discard everything except the cow. Though it sounds implausible and outrageous, this is not as destructive as our methods of deep sea fishing. These are tearing apart ecosystems that are 700 years old – the equivalent of the tropical forests in terms of richness of species – and leaving behind devastation. Unfortunately the wreckage is out of sight and for most of us it is not our concern.

It may come as a shock to realise that this horrendously destructive and inefficient way of catching fish is heavily subsidised by taxpayers' money. Norwegian taxpayers paid four million euro in subsidies to build Atlantic Dawn in a Norwegian shipyard. Globally, government direct or indirect aid to industrial fishing runs as high as $54 billion.[138]

Coral reefs

Shallow water habitats are also under threat. Coral reefs are the tropical forests of the oceans. Many have developed over hundreds of years and are replete with an extraordinary array of marine life. The wonderful structure of the corals themselves provides animals and fish with a safe habitat in which to breed and to feed. During my twenty years in the Philippines in the 1970s and 1980s I spent many hours snorkelling around reefs. I came to love the magic of coral reefs, their magnificent architecture and the pastel colours of the reef fish as they dash in and out of the coral cover. Corals are truly a wonderland.

Today many reefs are being fished out for luxury fish like groupers and humphead rass. The problem is that they are being fished out at an unsustainable level. Martha Holmes discovered that 30,000 tons of reef fish pass through Hong Kong each year. As stocks in reefs in nearby countries are depleted, fishing boats have to seek out new coral reefs, often 3,000 miles away. Marine biologists are worried that much of the catch is now composed of juveniles. You do not need to be a marine biologist to realise that when the bulk of your catch are juveniles the end is in sight for that species. Harvesting fish at a critical point in the life cycle before they can reproduce jeopardises the future of the species. It is like killing the goose that lays the golden egg!

Fishing for juveniles is not the only problem facing coral reefs. Human activity is putting other pressures on corals all over the world. A 1997 study on coral reefs coordinated by the University of Hong Kong found that coral reefs around the world are in a lamentable state. Researchers checked 300 reefs in 30 countries and found that a mere 32 per cent of the reefs had living corals. This means that 68 per cent were barren or seriously degraded. The Caribbean had the lowest rate of living corals at 22 per cent. South East Asia was just a little better off with 30 per cent. Marine scientist Edgardo Gomez of the University of the Philippines estimates that 90 per cent of the Philippine's 34,000 sq km of reefs are dead or deteriorating.[139]

During my years in the Philippines I witnessed the widespread disappearance of corals through siltation from deforestation and monocrop, industrial agriculture. Corals live in symbiotic relationship with zooxanthellae that depend on sunlight for photosynthesis. The silt washing down from the hills and mountains as a result of deforestation interferes with photosynthesis and as a result the corals die.

Walking along beaches in the province of Misamis Occidental in the Phillipines, I often heard the loud explosion as dynamite was used to stun and kill fish. In some ways it is an efficient way of catching fish today. But it ensures that there will be no fish tomorrow and a single blast can wipe out a patch of coral and destroy something that has taken decades to build up. Dynamiting can also take its toll on the fishermen. The fisherman has to be very dexterous in lighting the fuse and timing the moment when he throws the bottle with the dynamite into the water. If he throws too soon the wick will quench. There is very little margin for error and unfortunately many lose a hand as the bottle with the dynamite blows up in their hands. On the beaches of Misamis Occidental I met many men who had lost one hand and a few who had lost both. I was very aware that many of these people took what appears to us as a foolhardy decision because they were living in dire poverty and needed food immediately. Some paid dearly for this by losing their hands.

Cyanide is also used to stun fish, especially for the aquarium trade to First World countries. When cyanide is squirted into crevices in the reef the fish that rush out in search of oxygen are easily caught. The fish do not die immediately but their internal organs begin to collapse. By this time unscrupulous merchants have sold off the fish for aquariums in Europe, Japan or the US. Within a few weeks the fish die but the aquarium owner may feel that their death was due to natural causes or was caused by faults in their fish tank. The corals themselves also die after been sprayed with the poison. Within a few weeks a fertile and beautiful ecosystem disintegrates and is covered with algae.

Mangrove forests

Mangrove forests are another ecosystem in the shallow oceans that is under threat. All that some people see is their twisted root systems, insects and especially mosquitoes. They also emit unpleasant odours so that many people, especially those in urban areas, are delighted to see them destroyed and replaced by fish ponds, houses or hotels.

The destruction of mangroves is a tragedy. Like the coral reefs they are an extraordinarily productive life system and they provide food and shelter for breeding. The intricate root system provides a nursery for young fish which, when they mature, will move out to the reefs and open ocean. Over the past thirty years millions of hectares of mangrove forests have been destroyed. Thailand lost 27 per cent of its mangroves, Malaysia 20 per cent, the Philippines 45 per cent and Indonesia 40 per cent. The expansion of fish farming in these tropical countries has contributed to the destruction of mangroves in recent decades as have buildings for tourist development. Mangroves protect coastlines in many tropical countries from storms and typhoons.

Oil spills

The oceans of the world are being destroyed by oil spills. It is difficult to estimate how much oil enters the sea each year. One estimate would put the dumping as high as 25 million barrels.[140] The damage is most obvious when oil tankers break-up at sea. In November 2002 the single hulled oil tanker *Prestige* broke up in the Atlantic and disgorged 70,000 tons of fuel oil into the Atlantic. The oil has already polluted the coast of Galicia in north western Spain and threatens to become an ecological disaster for birds and all forms of marine life. Fisheries in the area will also be very badly effected. The oil slick has touched some of the most sensitive marine ecosystems in Spain. It affected fish, birds, dolphins and shell fish. The shoreline between the Cape Finisterre and La Coruna is home to one of the largest European populations of black-legged kittiwakes. There are more than ninety species of fish, including eleven species of sharks, bottle-nosed dolphins and porpoises in the areas and it is renowned for its shellfish.[141]

The *Prestige* was just the latest in a long list of marine disasters involving oil tankers in recent decades. On 19 July 1979, the *Atlantic Express* and *Aegean Captain* collided off

Tobago in the Caribbean releasing 287,000 barrels of oil into the ocean. On 24 March 1989, the *Exxon Valdez* ran aground in Prince William Sound, Alaska, releasing 37,000 tons of oil and contaminating 1,300 miles of coastline. The Braer ran aground in a storm off the Shetlands releasing 85,000 tons of crude oil. Prawn and mussel fishing in the area was banned for years. Further south, the Sea Empress spilled 72,000 tones of crude in the sea polluting 125 miles of the Pembrokeshire coastline.[142] In addition, oil is released into the sea from tankers each year as they wash out their holds.

It is essential that human beings begin to recognise that the destruction of the oceans impoverishes the planet for all future generations. The main losers in the human community are 200 million small scale fishermen in Third World countries like those in Mauritania. These people have lived for generations off the catches they have made around their native shores. Fish has also helped to feed their communities and has often provided the main source of food, especially protein.

One way of preserving the oceans would be to protect sufficient sections from human predatory activity so that stocks of fish and other marine animals could regenerate. Obviously, productive areas like coral reefs, mangrove forests and estuaries ought to have pride of place.

Martha Holmes makes the point that when we realise that land animals like the giant pandas need protecting we try to do something to save them. Unfortunately we treat the oceans in a very different way.

As a result of modern modern fishing technology all the oceans are accessible to humans. Unless we wish to deplete them of fish within the next two decades, and deprive future generations, we need to designate areas of the oceans where fish and marine animals will not be preyed upon by humans. There the fish can grow, develop, reach maturity and replenish the stock.

Only a small proportion of the world's oceans, around one third of 1 per cent, is now designated a marine sanctuary. On a global scale this is equivalent to the size of South Africa. Even in some of these places fishing is still allowed. Only one ten thousandth of the surface of the ocean is protected from all forms of fishing. That is equivalent to the size of the Netherlands and is certainly not enough to protect the oceans and the creatures of the oceans.

There are a few areas in the world where protective measures have been put in place. A reserve was created off the north island of New Zealand twenty-five years ago. In the intervening years stocks have recovered and the area is now teeming with fish. Lobsters are also abundant and growing a lot bigger. This five kilometres long reserve now produces as many lobsters as a one hundred kilometres stretch of unprotected coastline. Fishermen are allowed to fish right up to the boundary. The lobsters are so abundant that they have spread into the surrounding waters. In this way the reserve replenishes the fish stocks. The spawning stocks are protected and therefore fishing can continue in a sustainable way. Another example comes from the Grand Cayman Islands. Twenty years ago queen conchs, spiny lobsters and large groupers had all but disappeared from the local reefs. Then the government imposed controls on fishing and boating in the area. In the intervening years the conchs, lobsters and groupers have staged a remarkable recovery.[143] Professor Callum Roberts, a marine biologist at the University of York, insists that the only way to address the disappearance of fish in the oceans is to have a complete ban on fishing in many areas, especially where fish breed. He maintains that within these 'no fishing zones' fish will thrive and shoals will swim out into areas where they can be taken without doing damage to the breeding stock.

In May 2003 the Australian government released a draft plan to protect almost one third of the Great Barrier Reef from fishing and trawling. A representative section of each of the reef's seventy biological regions will be set aside. If the plan is carried through and acted on, it will make this area the largest protected marine area in the world. Opinion polls indicated that 80 per cent of respondents are willing to tolerate less access to the reef to give it more protection. At present, less than 5 per cent of the reef, which runs for over 2,000 kilometres is protected by its status as a national marine park.[144]

Given the present crisis that is looming, many experts feel that between ten and twenty per cent of our oceans ought to be designated no-take zones. We must learn to protect significant areas of different habitats – the deep sea, coral reefs, mangroves and spawning grounds. This is the only way that stocks will be able to recover in key fishing grounds around the world. Migration routes must also be protected. Britain has now declared the area of Lundy Island, off North Devon, as a no-fishing zone. The area is under the management of the National Trust.[145]

Education is also essential. Few people are aware of the plight of the oceans at the moment. The media has not set about informing people about what is happening to the seas in their own area or globally. In many circumstances there is a business-as-usual approach. The *Irish Times* marine correspondent Lorna Siggins wrote an article about the annual review of An Bórd Iascaigh Mhara (BIM). The review states that there has been a 10 per cent increase in the value of seafood on the Irish market at the point of sale. The report does acknowledge that there are pressures on certain fish stocks, but it still feels that ambitious growth targets set for 2006 can be achieved.[146] A reader would finish this article without any sense that the oceans are in crisis.

One of the best pieces of maritime news in recent times was the commissioning of the research vehicle RV *Celtic Explorer* in Galway Bay on 11 April 2003. The 210 feet long research ship will be able to undertake research in the 2,220 million acres of Irish maritime waters. Working together with Scottish and French research ships, scientists will be able to give an overview of the state and distribution of fish stocks from the Shetland Islands to the Bay of Biscay. These surveys can then be used to determine what catches are sustainable and what measures need to be taken to renew stocks.[147]

The National University of Ireland in Galway has a long tradition of involvement in marine science. Their capacity has been further enhanced by the building and funding of the Martin Ryan Institute (MRI) in recent years. According to the Director of the Institute, Dr Michael Guiry,

> Our vision for the institute is the careful use of traditional and new-age science to foster the balanced use of our marine resources, both physical and biological, for the benefit of our maritime communities.[148]

One would also hope that the Institute might also help the broader Irish community understand and respond to the challenges that are facing marine life at this point in time. Promoting marine sanctuaries in particular areas of Irish coastal waters might be one way to help stocks recover.

Water and the Christian Churches

The oceans

Because the Israelites were not seafaring people, like the Phoenicians or the Vikings, the oceans get very little attention in the Bible. In general the deep sea was perceived as a dark, mysterious and dangerous place.

At the beginning of the book of Genesis (Gen 1:1-2), we find 'the void' and 'the deep' and God's spirit hovering over the water. In the Babylonian creation myth *Enuma Elish* the original substance of the world was water. It was personified by the goddess Tiamat. All the gods were born from the union of Tiamat and Apsu. One of the gods, Marduk, went to war against Tiamat and eventually killed her. He splits her body in two and thereby formed the upper region – the sky – and the lower region – the earth. The idea behind this myth was that creation brought order to the world, particularly by bringing into check the primordial forces of nature, especially those associated with the oceans. The author of Genesis would have been aware of this tradition. He too sees creation as bringing order into being. But he did not a need a pantheon of gods and goddesses to do this. For him only Yahweh had the power to tame the elements, especially the unpredictable oceans. Taming the sea runs through many of the books of the Hebrew Bible. Yahweh challenges Job:

> Who pent up the sea behind closed doors when it
>   leaped tumultuous out of the womb,
> when I wrapped it in a robe of mist –
> and made black clouds its swaddling bands;
> when I marked the bounds it was not to cross
> and made it fast with a bolted gate

Come thus far and no farther;
Here your proud waves shall break.

(Job 38:8-11)

The need to keep the waters in check is also a theme of the psalms. The psalmist praises Yahweh's creative powers. These are evident in the fact that:

At your reproof the waters took to flight,
They fled at the sound of your thunder,
Cascading over the mountains, into the valleys,
Down to the reservoir you made for them;
You imposed the limits they must never cross again,
So they would once more flood the land.

(Ps 104:7-8)

The fearsome nature of the ocean and the dangers facing seafarers riding in fragile boats is also emphasised.

He spoke and raised a gale, lashing up towering waves.
Flung to the sky, then plunged to the depths, they lost
their nerve in the ordeal, staggering like drunkards
with all their seamanship adrift

(Ps 107:25-27).

In Psalm 69:14b-15 the psalmist implores God not to abandon him. The image he uses is the expansive ocean.

Save me from deep water!
do not let the waves wash over me,
do not let the deep swallow me
or the pit close its mouth on me.

Jonah's trip from Joppa to Tarshish, the lone sea journey in the Old Testament, reinforces this negative image of the sea as a dangerous place, possibly not far from the gates of the underworld (Jn 2:7). Even in the New Testament the sea is

presented as a dangerous place. The demons that terrorised the
Gerasene demoniac beseeched Jesus to allow them to enter the
swine who then 'charged over the cliff into the lake, and there
they were drowned' (Mk 5:13).

There is, of course, a more positive approach to the oceans
in the Bible. The oceans are created by God (Gen 1:9-10).

> God said 'Let the waters under heaven come together
> into a single mass and let dry land appear'. And so it
> was. God called the dry land 'earth' and the mass of
> waters 'seas', and God saw it was good.

Because they are God's creatures the Psalmist invites the
'oceans and all that move in them' to praise God (Ps 69:34). We
find the same theme in the Song of the Three Young Men, 'seas
and rivers! Bless the Lord' (Dan 3:78).

The Bible is clear that God's power transcends everything
even the might of the ocean.

> Yahweh, the rivers raise,
> the rivers raise their voices,
> the rivers raise their thunders
>
> Greater than the voice of the ocean,
> transcending the waves of the sea,
> Yahweh reigns transcendent in the heights
>
> (Ps 93:2b-4)

As Christians living in a world where the oceans are under
threat from human activity, we need to develop this positive
strand in the biblical teaching in order to shape a theology of the
oceans which will help us to protect the seas in our modern world.
The story of planet earth, or more correctly, planet water, must
become the basis for a new understanding of the importance of
the oceans for the well-being of the planet. Poisoned, polluted
oceans will bring illness and harm to every creature, including

humans. Stories and myths of various maritime people can also help develop a more caring attitude towards the oceans.

Fresh water in the Bible and the Church

Water has a central role in the Bible. It is there at the beginning of the creation of the world.

> In the beginning God created the heavens and the earth. Now the earth was a formless void, there was darkness over the deep, and God's spirit hovered over the water.
>
> (Gen 1:1-2)

Water is also present at the creation of humankind. In the second account of creation, just before the creation of Adam, 'a flood was rising from the earth and watering all the surface of the soil'. (Gen 2:6)

Yahweh planted a garden for human kind and 'a stream flowed from Eden to water the garden'. (Gen 2:10)

Later on in the Bible Jeremiah uses the metaphor of water to describe Yahweh as the source of life and he chides the people of Israel for turning their back on Yahweh and for attempting to find sustenance elsewhere.

> Since my people have committed a double crime: they have abandoned me, the fountain of living water, only to dig cisterns for themselves, leaky cisterns that hold no water.
>
> (Jer 3:13)

In Israel there was a very clear distinction between 'living water' and other kinds of water like that found in cisterns. Living water gushed from springs or was found in the flowing waters of the river. It was seen as fresh and not dependent on humans beings for storage. Many writers in the Bible, especially the prophets, apply this image of living water to Yahweh. He is the source of living water.

Hope of Israel, Yahweh! All who abandon you will be put to shame, those who turn from you will be uprooted from the land, since they have abandoned the living water.

(Jer 17:13)

In John's gospel in the New Testament Jesus applies this expression to Himself on a number of occasions.

In response to the surprise of the Samaritan woman who was taken aback when Jesus asked her for a drink he said:

If you only knew what God is offering
And who it is that is saying to you:
'Give me a drink',
you would have been the one to ask,
and he would have given you living water.

(Jn 4:10)

In reply the Samaritan woman expresses surprise at how Jesus, who has no bucket, could give her water. She also wonders whether Jesus is on par with Jacob, the patriarch, who gave them the well and used it himself. Jesus replied.

Whoever drinks this water
will get thirsty again;
but anyone who drinks of the water that I shall give
will never be thirsty again:
the water that I shall give
will turn into a spring inside him,
welling up to eternal life

(Jn 4:10,13,14).

Later in John's Gospel, on the last day of the festival, Jesus stood there and cried out:

If any man is thirsty, let him come to me!
Let the man come and drink who believes in me!

As scripture says:

> From his breast shall flow fountains of living water. He
> was speaking of the Spirit which those who believe in
> him were to receive for there was no Spirit as yet
> because Jesus had not been glorified.

> (Jn 7:37-39)

Water is also seen as a source of health and abundance. One
of the most powerful presentations of this view in the Hebrew
Scriptures is found in Ezekiel 47:1-12. It is a vision of
fruitfulness, abundance, and extolling the healing and life-
giving qualities of *clean* water. The prophet did not understand
the role of marshes and mudflats in marine ecosystems. Today
we know that without these ecosystems there would not be
such an abundance of marine life!

The prophet sees water pouring out from beneath the
Temple. Initially it reached his ankles, then his knees and waist,
welling up into a river that could not be crossed. The river

> ... flowed into the sea and made its waters wholesome.
> Wherever the river flows, all living creatures teeming
> in it will live. Fish will be very plentiful, for wherever
> the water goes it brings health... There will be
> fishermen on its banks. Fishing nets will be spread
> from En-gedi to Eneglaim. The fish will be as varied
> and as plentiful as the fish of the Great Sea
> [Mediterranean]. The marshes and lagoons, however,
> will not become wholesome, but will remain salt.
> Along the river, on either bank, will grow every kind
> of fruit tree with leaves that never wither and fruits
> that never fail; they will bear new fruit every month,
> because this water comes from the sanctuary. And
> their fruit will be good to eat and the leaves medicinal.

> (verses 9-12)

In the Book of Exodus, water is seen as a source of Israel's liberation from slavery in Egypt. Early in the book water played an important role in protecting the young Moses from being killed like other Jewish children because they were seen as a threat by the Pharoah. When Moses' mother could no long hide her baby from the Egyptians, 'she got a papyrus basket for him; coating it with bitumen and pitch. She put the child inside and laid it among the reeds at the river's edge'. (Ex 2:3-4)

During the exodus itself the Israelites were terrified that the pursuing Egyptian army would overtake them and destroy them. Yahweh ordered Moses

> Raise your staff and stretch out your hand over the sea and part it for the sons of Israel to walk through the sea on dry ground. I for my part will make the hearts of the Egyptians so stubborn that they will follow them. (Ex 14:15-17)

> Yahweh drove back the sea with a strong easterly wind all night, and he made dry land of the sea. The waters parted and the sons of Israel went on dry ground right into the sea, wall of water to the right and to the left of them. The Egyptians gave chase; after them they went, right into the sea, all Pharoah's horses, his chariots and his horsemen. In the morning watch, Yahweh looked down on the army of the Egyptians from the pillar of fire and cloud, and threw they army into confusion. He so clogged the chariot wheels that they could scarcely make headway. 'Let us flee,' the Egyptians cried, 'Yahweh is fighting for them against the Egyptians!' 'Stretch out your hand over the sea,' Yahweh said to Moses, 'that the waters may flow back on the Egyptians and their chariots and their horsemen'. Moses stretched out his hand over the sea. The returning waters overwhelmed the chariots and the horsemen of Pharaoh's whole army, which had followed the Israelites into the sea; not a single one was left. But the sons of Israel had marched through

the sea on dry ground, walls of water to the right and
to the left of them. That day, Yahweh rescued Israel
from the Egyptians, and Israel saw the Egyptians lying
dead on the shore. Israel witnessed the great act that
Yahweh had performed against the Egyptians, and the
people venerated Yahweh; they put their faith in
Yahweh and in Moses, his servant.

(Ex 14:21b-31)

The liberation of Israel by Yahweh was the bed-rock and core
of Israel's faith in Yahweh's great deeds that brought them
liberation. It was the fulcrum on which the whole life of the people
revolved. They celebrated the exodus each year during the paschal
meal. Through this celebration each generation of Israelites could
claim that they have also been liberated from Egypt.

In the description of the flood, water was seen as an
instrument of punishment.

For my part I mean to bring a flood and send the waters
over the earth, to destroy all flesh on it, every living
creature under heaven; everything on earth shall perish.

(Gen 6:7)

Noah is instructed to build an ark so that he, his family,
and two of each living creature might be saved from
drowning. 'Yahweh destroyed every living thing on the face
of the earth, man and animals, reptiles, and the birds of
heaven'. (Gen 7:23)

After the flood Yahweh made a covenant with Noah, his
family and every creature. 'No thing of flesh will be swept
away again by the waters of the flood. There shall be no flood
to destroy the earth again'. (Gen 9:11)

In the Hebrew Scriptures water also played a central role in
rites of purification. In the Book of Numbers we find a detailed
account of how to prepare lustral water in which ashes from a red
heifer are mixed. This water is used by people who have become
unclean by touching a corpse. Even the person who sprinkles the
lustral water must wash his clothing (Num 19:1-22).

The priest Aaron and his sons were obliged to wash their hands and feet before they entered the Tent of Meeting for fear they would die. This ordinance was for Aaron and his descendants from generation to generation. (Ex 30:18-21)

The same insistence on washing and cleaning with water is found in Leviticus. If a bronze vessel has been used for cooking sacrificial meat, 'it must be scrubbed and thoroughly rinsed with water' (Lev 6:22).

The beginning of Leviticus 15 obliges any male with a sexual impurity to wash both himself and his clothes. Those who come into intimate contact with him must also wash to become undefiled. (Lev 15:1-5)

Given this insistence on cleansing with water, it is easy to understand how water rituals were seen as cleansing people from sin. Mark, in Chapter 1 of his gospel, describes the ministry of John the Baptist:

> He appeared in the wilderness, proclaiming a baptism of repentance for the forgiveness of sins. All Judaea and all the people of Jerusalem made their way to him and as they were baptised by him in the river Jordan they confessed their sins.
>
> (Mk 1:4-5)

Jesus himself came and was baptised by John in the Jordan.

In chapter 7, Mark addressed the issue of ritual washing. The Pharisees noticed that some of Jesus' disciples were eating without washing their hands as prescribed by tradition. They challenged Jesus about this apparent disrespect for the tradition of the elders. Jesus used the occasion to challenge the Pharisees and others who will go to any length to maintain a tradition while they lack love, care and concern for others. He then went on to point out that uncleanness comes from inside a person and not from the outside

> Nothing that goes into a man from outside can make him unclean; it is the things that come out of a man

that make him unclean. If anyone has ears to hear, let
him listen to this.

(Mk 7:14-16)

Water has other symbolic meanings also in the Old
Testament. For the psalmist who had experienced the dry
parched lands of the Middle East water becomes a powerful
metaphor for an individual's longing for God. (Ps 42:1-2) The
plaintive cry of the exile is truly heart-rending.

As a doe longs
for running streams,
so longs my soul
for you my God.

My soul thirsts for God
the God of life;
when shall I go to see
the face of God?

One finds the same sentiments in Psalm 63:1-2.

God, you are my God, I am seeking you,
my soul is thirsting for you,
my flesh is longing for you,
a land parched, weary and waterless;
I long to gaze on you in the Sanctuary,
and to see your power and glory.

The Psalmist also sees water as a gift from God and essential
for life and food. In Psalm 65:9-10 the Psalmist gives thanks to
God.

You visit the earth and water it,
you load it with riches;
God's rivers brim with water
to provide their grain.

This is how you provide it;
by drenching its furrows, by levelling its ridges,
by softening it with showers, by blessing the first-
    fruits.

Water is used to describe the good life of the virtuous person: 'The mouth of the virtuous is a life-giving fountain, violence lurks in the mouth of the wicked'. (Proverbs 10:11)

For people who lived in the parched land of Palestine it is no wonder that plentiful, living water would be seen as one of the blessings that would accompany the dawn of the Messianic era.

A king shall reign by integrity
and princes rule by law;
each is like a shelter from the wind,
A refuge from the storm,
like streams of water in dry places,
like the shade of a great rock in a thirsty land.

(Is 32:1-2)

I will make rivers well up on barren heights,
and fountains in the midst of valleys;
turn wilderness into a lake,
and dry ground into water spring.

(Is 41:18)

The bounty of the new age will touch all creatures. All living creatures, as well as humans, will praise Yahweh because he provides for them with such loving-kindness.

No need to recall the past,
No need to think about what was done before,
See, in am doing a new deed,
Even now it comes to light; can you not see it?
Yes, I am making a road in the wilderness,
Paths in the wilds.

The wild beast will honour me,
Jackals and ostriches,
Because I am putting water in the wilderness
(rivers in the wild)
to give my chosen people drink,
the people I have formed for myself
will sing my praises

(Is 43:18-20).

The metaphor of water has other nuances. It is also used to capture the fickleness and infidelity of the people of Israel in their response to God's love. This is expressed in a most poignant way by the prophet Hosea. He asks:

What am I to do with you, Ephraim?
what am I to do with you, Judah?
this love of yours is like a morning cloud,
like the dew that quickly disappears.

(Hos 6:4)

In Isaiah 55:1-2 the poor will have access to pure, clean water when God's Kingdom is established. All are invited, even the poor who have no money:

Oh, come to the water all you who are thirsty;
though you have no money, come!
buy corn with money, and eat,
and, at no cost, wine and milk,
Why spend money on what is not bread,
your wages on what fails to satisfy?

In the New Testament Christ's own baptism in the Jordan is linked to his mission to bring about justice and peace for all: 'In the waters of the Jordan your Son was baptised by John and anointed by the Spirit'. Fr Killian McDonnell described the cosmic dimension of the baptism of Jesus. He recalls the writings of Gregory Nazianzus who has Jesus 'carrying the cosmos with him as he ascends out of the water of the Jordan'. He goes on to argue that

> the cosmic dimensions of the baptism of Jesus are part of antiquity's broader conviction, rooted in incarnation and resurrection, that the material universe, as the home of a redeemed humanity, is destined for transfiguration through the power of the Spirit manifested in the risen body of Christ.[149]

Based on our belief in baptism McDonnell argues that

> the ecological movement should have as its goal not only the preservation and restoration of the natural environment because we live and die here. Creation should be worthy of its vocation to praise. Praise him, sun and moon. Praise the Lord mountains and all hills, fruit trees and all cedars! Wild animals and all cattle, creeping things and flying birds (Ps 148). The cosmos lives in hope. The Universe is destined for God and for transformation.[150]

Water was important in Jesus' teaching mission. The incident with the Samaritan woman at the well in St John's gospel afforded Jesus the opportunity to present his life-giving message for all people. 'Whoever drinks this water will thirst again; but anyone who drinks the water that I shall give will never be thirsty again. The water I shall give will turn into a spring inside him, welling up to eternal life'. (Jn 4:14)

In Jesus' ministry, water was associated with cleaning and healing, In healing the blind man in John 9,

> Jesus spat on the ground, made a paste with the spittle, put this over the eyes of the blind man and said to him. Go and wash in the pond of Siloam (A name that means sent). So the blind man went off and washed himself, and came away with his sight restored. (Jn 9:6-7)

Baptising with water was to be a sign of new life and entry into His community. Jesus tells Nicodemus: 'I tell you most solemnly, unless a man is born through water and the Spirit he cannot enter the Kingdom of God' (Jn 3:5).

After his resurrection he told his disciples: 'All authority in heaven and earth has been given to me. Go, therefore, make disciples of all the nations; baptise them in the name of the Father and of the Son and of the Holy Spirit' (Mt 28:18-19).

We see in Romans 6:3-4 that in the early Church the celebration of baptism involved going down into the waters. This symbolised death and new life.

> When we are baptised in Christ Jesus we were baptised into his death; in other words, when we were baptised we went into the tomb with him and joined him in death, so that as Christ was raised from the dead by the Father's glory, we too might live a new life
>
> (Rom 6:3-4).

In the Christian tradition, the multiple symbolic functions played by water are highlighted and brought together in the blessing of the baptismal water in the Catholic Rite of Baptism. Water is the source of life. The prayer states that at the very beginning of creation:

> Your Spirit breathed on the waters, making them the wellspring of all holiness. The waters of the great flood you made a sign of the waters of Baptism, that make an end of sin and a new beginning of goodness. Water is also a sign of liberation, through the waters of the Red Sea you led Israel out of slavery, to be an image of God's holy people, set free from sin by baptism.

In a world where water is being polluted and abused one could argue that the symbolic connection between living water and the power of the Holy Spirit to incorporate those who are baptised into the Body of Christ is being compromised in a significant way. Can baptism with industrial or polluted water be considered vivifying, invigorating and life-giving?

Water is important even at the end of time. The vision of Ezekiel is recalled with the belief that when reconciliation and restoration take place in Christ in the New Jerusalem, living, clear and clean water will be abundant and sweet: 'Then the angel showed me the river of life, rising from the throne of God and of the Lamb, and flowing crystal-clear down the middle of the city street' (Rev 22:1).

The Church founded by Jesus is meant to be a community of service. Jesus highlighted this aspect of the community at the Last Supper by washing the feet of his disciples.

> When he had washed their feet and put on his clothes again he went back to the table, 'Do you understand,' he said 'what I have done to you? You call me Master and Lord, and rightly; so I am. If I, then, the Lord and Master, have washed your feet, you should wash each other's feet. I have given you an example so that you may copy what I have done to you'. (Jn 13:12-15)

The challenges of the contemporary world call Christians to serve not just the poor or those who are exploited or oppressed and have no voice. It also calls us to protect this fragile, water planet – earth.

Conclusion

Life began in water and can only survive if it has access to water. Modern science tells us that life emerged in the oceans about 3.8 billion years ago. For almost 2 billion years life remained there and evolved in extraordinary ways. Even when life came ashore, about 650 million years ago, it brought water with it in the cells of every living creature. We humans are almost 70 per cent water. Clean water has sustained and nourished life thoughout that magnificent journey from single celled organisms in the oceans to the complexity of human life and all other species on the planet. So, naturally, water is the most potent symbol of life. Water is a precious gift which we know to be under threat right around the world to day.

At the beginning of the chapter on 'Water in the Christian Churches' I mentioned that if the Vikings or Phoenicians had written the Bible there would have been a much more positive appreciation of the oceans. Even though the Irish are an island people, the ocean does not play an important role in our mythology or traditions – with one great exception, that is, of course, St Brendan's 'Journey to the Promised Land'. This ninth-century saga of the travels and exploits of St Brendan was translated into many languages and was retold around the fireside in villages, towns and cities on the coastlines of Europe for almost 1,000 years

Since the success of Tim Severin's voyage to America in 1977, by way of the Faroes, Iceland and Greenland, in a boat built according to the specifications found in *Navigatio*, many people now believe that St Brendan reached America almost 1,000 years before Columbus. The story combines accurate geographical information with some fanciful embellishments. There is a sense of intimacy with the ocean and the creatures of the ocean like the whale *Jasconius*. This amiable creature was willing to become an altar on each Easter Sunday so that the monks could offer mass on the whale's back. The whale did not become hostile when a fire was lit on his back. In fact, on Brendan's last visit, the whale took the monks on what today would constitute a farewell cruise. During the last few centuries St Patrick, though not a native of Ireland, has become the best known Irish saint. From the seventh to the sixteenth century St Brendan the Navigator was by far the best known and loved Irish saint. Maybe Irish people today could invoke his memory in order to develop a much more caring and sensitive approach to our oceans which we know are very much in peril.

Religious respect for fresh water in Ireland, especially holy wells, goes back to pre-Christian times. The Celts recognised water as the first and principal source of life. Many of the major rivers of Europe, like the Seine, the Rhine, the Marne, the Clyde, the Severn and the Shannon, still have Celtic names. The source of rivers inspired special veneration, as can be seen from archaeological remains all over what was once Celtic Europe. These rivers were associated with various fertility goddesses.

Sacred wells like the great healing centre in Bath in England and local springs were placed under the protection of the mother goddess.[151] They were places of pilgrimage associated with healing. The arrival of Christianity in the fifth century did not lead to an abandonment of holy wells. In fact they were 'Christianised' and often associated with a local saint, like Senan, Declan or Brendan. The faithful went on pilgrimage to the well to perform what was called the 'pattern'. This involved the recitation of certain prayers and the performance of various rituals. Almost every parish in Ireland has at least one holy well. Charles Plummer in his study of the *Lives of the Irish Saints* estimated that there were about 3,000 holy wells in Ireland.[152]

In post-Reformation Ireland, both the State and the Established Church viewed the 'pattern' as superstitious and attempted to suppress it. Even members of the Catholic hierarchy were opposed to it. They alleged that patterns were occasions for drunkenness and faction fighting. Still, they survived right into the twentieth century in many parts of the country.

With the growing consciousness of the need to protect our water resources, each parish ought to revisit and, in some situations, rediscover its holy well. The quality of water in the holy well could become a barometer for the quality of groundwater in the locality. If initially the water quality was poor the community could then begin to take the necessary steps to improve the water situation by addressing the problem in each household, farm or industrial enterprise. The aim of each community ought to be to restore the water quality of their holy well to such a level that it could be used in the sacrament of baptism. Then it would truly be life-giving water.

In his Lenten Pastoral 2003, Dr Dermot Clifford, Archbishop of Cashel and Emly, wondered whether an environmental competition between communities modelled on the Tidy Towns competition might help make people more environmentally aware. My suggestion is that a competition based on the purity of the water in the local holy well would be the place to begin. Caring for the well could give rise to a new element in the pattern pilgrimage. It would not be confined to the search for personal sanctity but would also include a commitment to social justice and the integrity of creation.

Christian leaders in Ireland and elsewhere would also do well to follow the leadership given by the Ecumenical Patriarch Bartholomew 1, the senior leader of the Orthodox Church. In 1997 he and a number of religious leaders, scientists, and environmentalists took a boat trip around the Black Sea, one of the most polluted seas in the world, visiting all six countries whose coastline it touches. They saw at first hand the damage that humans are doing to the planet. This once rich sea has been devastated by agricultural and industrial waste which has accumulated over the past fifty years. During the trip – a floating seminar – the participants heard from scientists who presented the latest data on the crises facing the Black Sea, and from theologians who spoke about how particular aspects of Christian faith related to the environment. The purpose of that floating seminar was to identify the environmental problems in each locality and to signal to Christians that they have an important stewardship role to play in both healing and protecting the planet.

In 2002 the floating seminar toured the Adriatic Sea. Once again protection of the sea was at the top of the agenda. When the ship docked in Venice on the 12 June 2002 Patriarch Bartholomew and Pope John Paul 11 signed a Declaration on the Environment. This document is the latest indication that concern for the degradation of the planet is now firmly established as one of the Churches' major concerns in the contemporary world. Such initiatives by religious leaders might rouse politicians to tackle the various water issues which I have highlighted in this book. The *World Water Development Report* published in 2003 was made possible by unprecedented collaboration between twenty-three different United Nations agencies. The Report states categorically that the world is facing a serious water crisis and unless it is tackled it will get much worse. It criticises the lack of commitment on the part of politicians at a national and international level to respond to the crisis in a comprehensive and effective way. The document states that the world is facing its worst human crisis with 'inertia at the leadership level and a world population not fully aware of the scale of the problem'.[153]

If religious leaders around the world followed the example of Patriarch Bartholomew then political leaders might be more willing to engage this difficult issue. In North Tipperary a floating seminar on Lough Derg would see Christian leaders from Clare, Tipperary, Limerick and Galway, along with scientists, farmers, personnel from the EPA and members of NGOs like Save Our Lough Derg reflecting on the quality of the

water in the lake. Such a seminar would help to bring ethical
and religious underpinning to the policy decisions that are
needed if Lough Derg is to be returned to the unpolluted state
I remember when growing up in Nenagh in the 1950s. If this
and similar initiatives were to take place around the country
then it would be clear to everyone that the ecological crisis is
now at the heart of Christian pastoral ministry and that it is
being addressed in an ecumenical way.

There are no Protestant lakes or Catholic rivers. Working
together as Christians to protect the environment could build
bridges of trust between the various Churches in Ireland that
have been divided by historical, political and cultural factors.
Unfortunately, at the moment the environment is not
anywhere near the top of the pastoral agenda in any of the
Christian Churches. Sadly when it is mentioned at all, it is
merely an addendum.

The reality is that destruction of the earth is one of the most
pressing challenges facing humankind. The stakes are very
high – a safe, beautiful and bountiful planet for future
generations or a degraded and barren one. To ensure that the
first option is achieved, people of every religion, but
particularly Christians, are called to dedicate their lives in
service of the world and the poor of the world. Working to
protect water and make sure that it is available freely to every
one on the planet is following in the way of Jesus in our world
today.

Notes

. . . .
. . . .

1   Pope John Paul 11, 'God made man the steward of creation', *l'Osservatore Romano*, 24 January 2001, 11 (unfortunately the quotation uses non-inclusive language).

2   'The party is over', *New Scientist*, 7 September 2002, p. 3.

3   *Global Environment Outlook 3*, London: Earthscan Publications, 2002. (www.earthscan.co.uk)

4   Patrick Smyth, 'The great water divide', *Irish Times*, 22 March 2003, 6.

5   John Vidal, 'Blue Gold: Earth's Liquid Asset', *Guardian (Earth)*, August 2002, 6.

6   Michael S. Serrill, 'Well Running Dry', Special Issue of *Time* (Our Precious Planet), November 1997, 19.

7   'Needing and Getting', *New Internationalist*, March 2003, 19.

8   Patrick Smyth, op. cit., *Irish Times*, 22 March 2003.

9   *Earth*, Health-check for a planet and people under pressure, *The Guardian*, in association with Action Aid, August 2002

10   Archbishop Renato R. Martino, www.vatican.va/roman_curia/po../repcjustpeace.doc_kyoto_wat er4/29/03.687

11   Iva Pocock, 'Clearly a better glass of water', *Irish Times*, 21 January 2003, 15.

12   Patrick Smyth, op. cit., *Irish Times*, 22 March 2003.

13   'Solution on Tap', *Guardian*, 10 July 2003.

14   Paul Brown, 'Failure to manage water kills two million a year – UN', *Guardian*, 11 April 2002.

15   'Water Filter Set to Save Lives', *EarthWatch*, Winter 02/03, 33.

16   Dinyar Godrej, op. cit., 'On the challenge posed by the world's freshwater crisis', *New Internationalist*, March 2003, 11.

17   Ibid., 10.

18   John Vidal, op. cit., *Guardian (Earth)*, August 2002.

19   Dinyar Godrej, op. cit., 11.

20   Payal Sampat, 'Groundwater Shock', *WorldWatch*, January/February 2000, 10.

21   Ibid., 14-15.

22   Ibid., 18.

23   Ibid., 20.

24 Paul Brown, 'Contamination – gender-bender chemicals are now inside all of us', *Guardian*, 12 January 2000, Supplement 4.

25 Don Hinrichsen, 'A Human Thirst', *WorldWatch*, Jan/Feb 2003, 15.

26 Polychlorinated biphenyls (PCB) are very stable compounds and can only be destroyed through incineration at very high temperatures. They have been used in a wide range of products from flourescent light bulbs to hydraulic fluid and most important of all, electric transformers. Their toxicity was first recognised in the 1930s but it was not until the late 1960s that the dangers posed by PCBs became widely appreciated. They have been banned in many First World countries.

27 Michael S. Serrill, op. cit., *Time*, November 1997.

28 Ibid., 14-15.

29 Dr William Reville, 'Water, water everywhere, but not for everyone', *Irish Times*, 15 May 2000, 9.

30 Philip Ball, 'Running on empty', *Financial Times Weekend*, 2 & 3 October 1999, 10.

31 Isaac, Jad, 'Water Conflict in the Holy Land', *Thinking Mission*, published by the United Society for the Propagation of the Gospel, Partnership House, 157 Waterloo Road, London SE1 8XA, 15-19.

32 John Burton, 'Malaysia puts the screw on Singapore over water', *Financial Times*, 7 March 2002, 8.

33 Sandra Postel, 'The Politics of Water', *WorldWatch*, July/August 1993, 14.

34 Paul Brown, op. cit., *Guardian*, 11 April 2002.

35 Fred Pearce, 'Water War: India could suck Pakistan dry', *New Scientist*, 18 May 2002, 18.

36 Fred Pearce 'Conflict looms over India's river plan', *New Scientist*, 1 March 2003, 4.

37 www.nytimes.com 2003/01/05, 'Today's Headlines'.

38 'Future looks bleak for Iraq's fragile environment', *New Scientist*, 15 March 2003, 12.

39 Don Hinrichsen, op. cit., WorldWatch, Jan/Feb 2003, 15.

40 www.nytimes.com 2003/01/11 'Politics'.

41 'Editorial', *Irish Times*, 14 March 2000

42 Donal Hickey, *Irish Examiner*, 28 August 2000, 15.

43 Frank McDonald, 'Implementing directive on water pollution poses major challenge', *Irish Times*, 5 July 2002, 6.

44 Treacy Hogan, 'One-third of rivers poisoned, reveals new water survey', *Irish Independent*, 29 December 2001, 9. See also *Water Quality in Ireland 1998-2000*, Environmental Protection

Agency, PO Box 3000, Johnstown Castle Estate, County Wexford, ix. (www.epa.ie)

45 *Water Quality in Ireland 1998-2000*, ix.

46 Dr Martin Knox is presently Managing Director of Amberley Quality and Environmental Services. Personal communication, 7 May 2003

47 'Water Ruling welcomed', *Irish Times*, 15 November 2002, 2.

48 National Rural Water Monitoring Committee, Report on Survey of Source Water Quality in Private Group Water Schemes, November 2000-2001, Executive Summary, July 2002, by M. J. O'Connell & Co, 18, 19.

49 'Bacteria found in Kilkenny holy well', *Irish Times*, 7 November 2002, 2.

50 Neans McSweeney, 'Concern over Kilkenny water quality', *Irish Examiner*, 4 April 2003, 2.

51 Frank McDonald, 'EPA criticised over water schemes', *Irish Times*, 2 May 2002, 2.

52 Ibid., 2.

53 Robert Pocock, 'Kilkenny only one of many counties at risk from aluminium and fluoride in drinking water', VOICE Press Release, 3 April 2003.

54 *A Scientific Critique of the Fluoridation Forum Report*, Ireland 2002. VOICE, October 2002, 5.

55 Kevin O'Sullivan, 'Danger of toxic microbes in Irish Lakes highlighted', *Irish Times*, 5 January 2000, 5.

56 Elena Bennett and Steve R. Carpenter, 'P soup', *Ecologist*, March/April, 2002, 26.

57 Ibid., 29.

58 *Water Quality in Ireland 1998-2000*, op. cit., ix.

59 David Santillo and Iva Pocock, 'Last chance to save Lough Sheelin', Greenpeace Exeter Research Laboratory, March 1995, 1.

60 Kathy Sheridan, 'EU water directive may save Sheelin', *Irish Times*, 3 May 2003, 2.

61 This is a device in which microbes break down the manure into biogas (methane) and digested solids. The methane can be used as a source of energy and the organic solids can be returned to agricultural land.

62 Mary O'Connor, 'The slurry worry', *Irish Times*, 9 August 2000, 7. Commercial Report Section.

63 Lloyd Gorman, 'Working to save troubled waters', *Irish Times* (A Commercial Report), 9 August 2000, 6.

64   Declan Fahey, 'Group warns on sewage threat to Shannon', *Irish Times*, 10 April 2002, 2.

65   Ibid., 2.

66   A Code of good Agricultural Practices to Protect Waters from Pollution by Nitrates, published in July 1996. It was developed by the Department of Agriculture, Food and Rural Development and the Department of the Environment and Local Government. It is promoted by Teagasc.

67   Tim O'Brien, 'EU advocate rejects Irish defence in water case', *Irish Times*, 27 April 2002.

68   Ibid.

69   *Water Quality in Ireland 1998–2000*, xv.

70   Ibid.

71   *The Water Framework Directive – issues and opportunities, An NGO Agenda for Ireland's Waters.* VOICE, 2003. www.voice.buz.org

72   Andrew Osborn and David Ward, 'Britain's towns top European pollution poll', *Guardian*, 20 March 2001, 7.

73   Paul Brown, '97 % of beaches make the grade for swimmers', *Guardian*, 26 April 2002, 6.

74   Dinyar Godrej, 'Precious fluid: challenge posed by the world's freshwater crisis', *New Internationalist*, March 2003, 10. www.watersave.com.au

75   Fr Charles Rue, 'The Gift of Water', *The Far East* (Melbourne) March 2003, 1.

76   The Center For Public Integrity, www.icij.org/dtaweb/water, Cholera and the Age of Water Barons, 1 April 2003.

77   Tremolet Sophie, 'Not a drop to spare', *Guardian*, 26 September 2001, 9.

78   The Center for Public Integrity, op. cit., 2.

79   Ibid., 4.

80   The Center for Public Integrity, op. cit., 2.

81   'Ways sought to avert conflicts over water', *Irish Times*, 22 March 2003.

82   'Public Statement by Amnesty International on Human Rights to Water', www.waterobservatory.org/new, Posted 27 March 2003.

83   Ibid., 'Critics accuse the forum of sidestepping the UN', Posted 27 March 2003.

84   Archbishop Renato R. Martino, 'Water An Essential Element for Life', http://www.vatican.va/roman_curia/po.. /re pc justpeace_doc_kyoto-water 4/29/03, 6 and 7.

85   Ibid., 10.

86   Curtis Runyan, 'Privatising Water', *WorldWatch*, January/February 2003, 36-38.

87   Ibid., 37.

88   Polly Ghazi, 'Running on empty', *Guardian* (Supplement), 31 March 1999, 4-5.

89   Ibid., 5.

90   Peter Gleick, 'World Water Spending Priorities Misguided', www.pacinStorg/kyoto, 03/04/03, 2, 4

91   Patrick, McCully, 'Big Dams, big trouble'. *New Internationalist*, March 2003, 15.

92   Tim Radford, 'China to pump rivers 800 miles north', *Guardian*, 27 November 2002, 16.

93   Editorial 'Cool, Clear Water', *International HeraldTribune*, 30 August 2002, 6.

94   Shahid Husain, 'Salt in wounds', *Guardian*, Supplement, 15 January 2003, 8-9.

95   Michael S. Serrill, 'Wells Running Dry', *Time*, November 1997, Special Issue, *Our Precious Planet*, 18.

96   Much of the data for this section on dams comes from Michael Beazley, *The Earth Report*, Beazley-International Limited, Artists House, 14-15 Manette Street, London WRV 5LB, 47-49.

97   Patrick McCully, 'Big dams, big trouble', *New Internationalist*, March 2003, 14.

98   This is another name for schistosomiasis. This disease is caused by flukes which have complex life-cycles involving specific fresh-water snail species as intermediate hosts. The infected snails release large numbers of free-floating larvae that are capable of penetrating human skin. The parasite is contracted in fresh water and can lead to fever and enlargement of spleen and liver.

99   Philip Ball, 'Running on empty', *Financial Times Weekend*, 2 & 3 October 1999, 10.

100 Patrick McCully, op. cit., *New Internationalist*, March 2003, 16.

101 Tim Radford, 'China to pump rivers 800 miles north', *Guardian*, 27 November 2003, 16.

102 Don Hinrichsen, 'A Human Thirst', *WorldWatch*, January/February 2003, 17.

103 Michael S. Serrill, op. cit., *Time*, November 1997, 21.

104 Fred Peirce, *New Scientist*, 7 September 2002, 7-8.

105 Ibid., 11.

106 John Vidal, op. cit., *Guardian (Earth)*, August 2002, 8.

107   Esther Addley, 'Tourists' water demands bleed resorts dry',
       Guardian, 12 May 2001, 13.

108   Ibid., 13.

109   Ibid., 13.

110   Ros Coward, 'Clear conflict', *Guardian*, Supplement, 27
       November 2002, 8-9.

111   Michael S. Serrill, op. cit., *Time*, November 1997, 19.

112   *Global Environment Outlook*, 162.

113   Ibid., 165.

114   *Our Common Future: The World Commission of Environment
       and Development*, 1987, Oxford University Press. This is a
       United Nations sponsored study on long-term environmental
       conservation and sustainable development, especially for the
       poor. It was chaired by Gro Harlem Brundtland who was Prime
       Minister of Norway at the time.

115   Ibid., 168.

116   William Reville, 'How the world's oceans cool us down', *Irish
       Times*, 22 May 2003, 13.

117   Kathy Brewis, *Sunday Times*, 10 June 2001, 32-38.

118   Ibid., 38.

119   *Global Environment Outlook 3*, London: Earthscan Publishing,
       2002, 181.

120   Don Hinrichsen, 1998, 'The Ocean Planet', *People and the
       Planet*, 8.

121   Ibid., 182-183.

122   Don Hinrichsen, op. cit., *WorldWatch*, Jan/Feb 2003, 6-7.

123   Ian Sample, 'Fall in fish stocks hits crisis point', *Guardian*, 15
       March 2003, 10.

124   William D. Montalbano, 'The Chips are down', *Guardian*, 15
       March 1995, 5.

125   Paul Brown, 'North Sea in crisis as skate dies out', *Guardian*,
       21 March 2002.

126   Harry McGee, 'No Codding, it could be the end for fishing',
       Sunday Tribune, 3 November 2002, 8.

127   Mark Hennessy, 'Fishing in Irish Sea may end after warning on
       cod stocks', *Irish Times*, 2 November 2002, 1.

128   Lorna Siggins, 'New Cod recovery plan criticised by industry',
       Irish Times, 8 May 2003, 2.

129   John Hearne, 'New species; fish out of water on Irish market?',
       Irish Examiner, Supplement, 25 March 2003, 5.

130   'Fishing For Trouble', *Ecologist*, April 2003, 18-19.

131 Dan Buckley, 'Africa's subsistence fishermen have little to thank Europe for', *Irish Examiner*, 6 February 2002, 19.

132 Ibid., 19.

133 Ibid., 19.

134 *Guardian*, February, 2003

135 Originally published in *Science*, February 1998, and quoted in Peter Mongatuge, 'Oceans without Fish', *Third World Resurgence*, April 1998, 5.

136 Lorna Siggins, 'Scientist criticises system of monitoring fish catches', *Irish Times*, 31 March 2003, 2.

137 Ibid., 2.

138 James O. Jackson, 'The Grim Sweepers', *Time*, Supplement on the State of the Planet, *Oceans*, 28 October 1996.

139 J. Madeleine Nash, 'Assault of the Reefs', *Time*, Supplement on the State of the Planet, Oceans, 28 October 1996,.

140 Bruce McKay and Kieran Mulvaney, 'Cleaning up the Seas', *People and the Planet*, 1998, 14.

141 David Sharrock, 'Rich ecosystem in jeopardy', *The Times*, 19 November 2002, 11.

142 'A sea of troubles', *Guardian*, 20 November 2002, 5.

143 J. Madeline Nash, 'Splashes of Hope', *Time*, 26 October 1996.

144 Stephanie Peatling, 'Plan to lock up 30 per cent of Barrier Reef', *New York Times*, 2 June 2003. www.shm.com.au/2003/01

145 Ian Sample, op. cit., *Guardian*, 15 March 2003, 10.

146 Lorna Siggins, 'Seafood industry reaches record levels of investment and sales', *Irish Times*, 29 March 2003, 18.

147 Lorna Siggins, '31 Million euro marine research vessel goes into service', *Irish Times*, 12 April 2003, 2.

148 Martin Ryan Institute, *Sunday Tribune*, 30 March 2003, 23.

149 Killian McDonnell, *The Baptism of Jesus in the Jordan*, Collegeville, Minnesota: Liturgical Press, 1996, 243.

150 Ibid., 244.

151 Sean McDonagh, *To Care for the Earth*, London: Chapman, 1985, 205.

152 Quoted in Patrick Logan's *The Holy Wells of Ireland*, Buckinghamshire: Colin Smythe, 1980

153 'Coming Soon: A Thirsty Years War', *The Times*, 31 July 2003, page 4-5 of the supplement.